# Vandas and Ascocendas

*Vanda sanderiana* 'Orchidgrove'

# Vandas and Ascocendas

## and Their Combinations with Other Genera

DAVID L. GROVE

*with paintings by*
ANGELA MIRRO

*and photographs by*
CHARLES MARDEN FITCH

Timber Press
Portland, Oregon

Timber Press, Inc.
The Haseltine Building
133 S.W. Second Avenue, Suite 450
Portland, Oregon 97204, U.S.A.

Printed in Hong Kong

Library of Congress Cataloging-in-Publication Data

Grove, David L. (David Lawrence), 1918–
    Vandas and Ascocendas and their combinations with other genera / David L. Grove ;
with paintings by Angela Mirro and photographs by Charles Marden Fitch.
        p.     cm.
    Includes bibliographical references (p.     ) and index.
    ISBN 0-88192-316-8
    1. Vanda. 2. Orchid culture. 3. Vanda—Breeding. 4. Hybridization, Vegetable.
I. Title.
SB409.8.V36G76  1995
635.9'3415—dc20
                                                                94-44663
                                                                CIP

# Contents

Preface    9

Acknowledgments    13

PART ONE   The Species and Their Hybrids

1  The *Vanda* Species
    Origin and Composition of the Genus    19
    Descriptions of *Vanda* Species    22

2  The Crucial Role of Chromosomes and Genes
    The Purpose and Process of Cellular Creation    38
    Some Practical Conclusions    45

3  The Species Ancestry of *Vanda* Hybrids
    The Determinants of Hybrid Ancestry    48
    Traits of *Vanda sanderiana* and *Vanda coerulea*    50
    The Contributions of Other Species    53
    The International Record of *Vanda* Hybrids    55

4  Criteria of *Vanda* Flower Quality
    The Need for Standards    71
    Judging Systems    72
    The Criteria for Judging    74

5  Ascocentrums and Ascocendas
    Origin and Composition of the Genus *Ascocentrum*    87
    Origin and Importance of the Genus *Ascocenda*    93
    The *Ascocentrum* Species Used in Making Ascocendas    94

The Lineage of the Award Winners     95
The Characteristics of Awarded Ascocendas     97

6   Other Intergeneric Combinations with *Vanda*
Introduction     99
Combinations with *Papilionanthe* and *Holcoglossom*     103
Some Combinations with *Arachnis*     104
Some Combinations with *Aerides*     109
Some Combinations with *Rhynchostylis*     114
Some Combinations with *Renanthera*     121
Some Combinations with *Paraphalaenopsis* and *Phalaenopsis*     125
Some Combinations with *Neofinetia*     131
Some Combinations with *Vandopsis*     132
Conclusion     134

PART TWO    Plant Selection and Care

7   A Guide to Buying Seedlings
General Considerations     137
Ancestry of Seedlings     140

8   Providing the Right Environment
Environment Versus Cultural Practices     143
Basic Environmental Requirements     145
Light     147
Temperature     160
Watering and Humidity     168
Air Circulation     173

9   Nutrition and Other Aspects of Plant Care
General Principles     176
Regimens for Proper Nutrition     178
Water Quality     184
Containers and Potting Media     188
Growing Small Seedlings     193
Topping     195
Pests and Diseases     196
Treatment of Stressed Plants     203

APPENDICES

A   A Guide to Orchid Classification and Nomenclature          207

B   Christenson's List of the Genus *Vanda*                    214

C   Score Sheet of the American Orchid Society for Judging Flower   218
    Quality of Vandas

D   Registered Intergeneric Combinations with Strap-Leaf Vandas   219


Bibliography                                                    223

INDICES

Botanical Names                                                 229

Biographical Names                                              236

Subject                                                         238

*Color plates follow pages 52 and 132*

# Preface

This book is about *Vanda* species and *Vanda* hybrids, and the many combinations of these with nine other related genera: *Aerides, Arachnis, Ascocentrum, Neofinetia, Paraphalaenopsis, Phalaenopsis, Renanthera, Rhynchostylis,* and *Vandopsis.*

The combination of *Vanda* species and hybrids with those of the genus *Ascocentrum* produces hybrids that are grouped into the man-made genus *Ascocenda,* and form that genus. Because *Ascocenda* has been by far the most numerous and popular of the intergeneric combinations with vandas, this man-made genus and its component genus *Ascocentrum* receive separate and more extensive coverage in the organization of this book.

Few other orchids have flowers with the breathtaking beauty, the wide range of rich colors, and the oftentimes crystalline texture of today's vandas and ascocendas and their combinations with other genera. The individual blossoms are attractively shaped and pleasingly presented on erect inflorescences. The plants are floriferous both in number of flowers and in frequency of flowering. Most of them bloom twice and sometimes three times a year. Their flowering period is not limited to any single season; they can bloom at any time of the year, and their flowers are long-lasting. Many bloom at an early age and on plants no more than 30 cm (12 in) tall. The showy inflorescences make fine cut flowers and the individual blossoms are excellent for corsages.

With reasonable attention to sanitation, the plants are not much troubled by insects or disease. They are far easier to maintain in good condition than are *Phalaenopsis,* a genus found in the orchid collections of nearly all hobbyists. Wherever cattleyas are being grown successfully, strap-leaf vandas, ascocendas, and some of their combinations with other genera also can be grown. While normally they need a greenhouse or a lathhouse, and generally do not lend themselves to windowsill or under-lights cultivation, the ability of most of them to adapt to sub-

9

optimal conditions of light and temperature is quite astonishing. Under subopti-mal conditions, the size, number, and intensity of color of the flowers will not dis-play the full potential of the plants, but they will bloom and give much pleasure nonetheless, provided they have a good genetic background.

Yet, despite their striking beauty and other outstanding attributes, these orchids are not as well known or appreciated as other popular genera. There are three rea-sons for this. First, the quality of most vandas and ascocendas before the mid-1980s left much to be desired; there have been vast improvements in the color and shape of their flowers since then—a brief period, given the time required to breed and bloom orchids. Second, there is a widespread misperception that these orchids are difficult to grow in temperate climates. Most orchid growers think that they all require very tropical conditions. And, third, little has been written about them in any depth. The aim of this book is to provide a better understanding of vandas and ascocendas, and of their combinations with related genera, enumerating their de-sirable qualities, discussing how these were achieved, and dispelling the notion that these orchids are more demanding in their cultural requirements than are the more commonly grown types. The book is divided into two main sections. Part One traces the evolution of today's *Vanda* and *Ascocenda* hybrids (including their intergeneric combinations with *Vanda*); Part Two provides a detailed reference manual for plant selection and successful culture. Parts One and Two are quite dif-ferent; each has its own objective, but each is designed to complement the other. Both are needed to fully appreciate the wonders of vandas and ascocendas and their hybrids with related genera.

Part One examines the history and evolution of today's *Vanda* and *Ascocenda* hybrids, their combinations with other compatible genera of orchids, and criteria for evaluating the results. The treatment of hybrids begins with a brief review of the genetic processes by which traits are transmitted to future generations and the genetic variability that provides the foundation on which selective breeding for im-provement depends. Hybridizers obviously had to start with the available species. What were they? Which ones did they use in their breeding programs? What traits were they trying to produce in the progeny? Was there a consensus about the ideal they were seeking, or was there a great divergence of goals? Has hybridizing un-dergone periods of fads or fashions? Which species appear on the ancestral tree of the best modern hybrids, and which ones thus far have failed to produce descen-dants considered worthy of further use in mainstream breeding programs? What are the main qualities—good and bad—of some of the species that were given chances to demonstrate what they could contribute to hybridizers' goals? What criteria are used to evaluate the flower quality of hybrids?

With an understanding of the qualities and genetic potential of vandas and as-cocendas and of their combinations with other genera, the pleasure of growing these exciting orchids will be enriched. Most readers, however, will not be satisfied with intellectual knowledge alone; their primary concerns probably are how to select plants and grow them well and, particularly, how to cope with daily cultural problems. Part Two of the book addresses those concerns in very specific terms. In the course of many talks I have given to orchid societies about vandas and ascocendas, one thing has stood out: growers have an insatiable appetite for the most minute details of plant selection and cultural practices, and especially for the "recipes" of skilled growers. There is no lack of readily available information about the generalities of good culture. It is the translation of those generalities into specific application that raises questions. Part Two provides growers with a reference manual to which to turn whenever a baffling problem arises with culture, environment, or greenhouse management.

Throughout the book, for convenience' sake, whenever the phrase "vandas and ascocendas" is used, it will be meant to include the other intergeneric combinations with *Vanda*, except where the context makes it obvious that the term is being employed in the narrow sense.

The use of certain taxonomic and nomenclatural terms is unavoidable in this book. Readers not well versed in the peculiarities of orchid classification and nomenclature are strongly urged to begin by reading the brief, and somewhat simplified, Guide to Orchid Classification and Nomenclature, which is presented as Appendix A.

# Acknowledgments

Two groups of individuals contributed greatly to this book: (1) expert orchid growers who, while in no way directly involved in the preparation of the manuscript, unstintingly shared their extensive knowledge, skills, and experience; and (2) other experts who reviewed parts or all of the manuscript.

In the first group is Pravit Chattalada, who, over a period of 20 years, spent uncountable days taking me to markets and nurseries throughout Thailand during my frequent trips to his country. He and his charming wife, Revadee, who often accompanied us on our excursions, especially during the early years, gave me my first tutorials on what to look for and how to evaluate what we encountered. It was they who introduced me to Charungraks Devahastin, a leading pioneer hybridizer of vandas in Thailand and owner of CD Orchids in Bangkok. Charungraks spent many long days taking me to "up-country" nurseries that no nonresident could possibly locate; these nurseries are the main source of the vandas and ascocendas that are exported. (In Thailand, in many cases there is a division of labor between breeding and growing.) For Charungraks, the "opportunity cost" of our trips was the sacrifice of innumerable games of golf, his favorite pastime. He always was unreluctant to share his cultural and hybridizing "secrets." It was a great shock to hear that he was killed in an automobile accident in April 1993.

Other growers and hybridizers in Thailand who were generous with their time and knowledge are Treekul Sophonsiri, owner of Kultana Orchids in Bangkok, and Kasem Boonchoo, also of Bangkok and the originator of several of the most influential *Vanda* and *Ascocenda* hybrids in the 1980s (some registered by him and some registered by others). Many other commercial growers and breeders too numerous to mention by name also were hospitable and free with their time, perhaps because of their friendship with Charungraks Devahastin or Pravit Chattalada.

In the United States, orchidists Marilyn Mirro and Robert Fuchs always have

13

been willing to share their experience whenever a special problem of culture has arisen; in the course of writing this book, I often felt the need to check some of my tentative impressions and conclusions with each of them. In addition, Bob Fuchs provided color photographs of some of his outstanding plants.

Charles Marden Fitch, the dean of horticultural photographers and author of several books, both on orchids and on other horticultural subjects, has permitted the use of many of his color pictures. William Smiles did the microphotography that reveals the unique structural characteristics of the lips of *Vanda coerulea* and *V. sanderiana*. He also provided some other photographic material.

Angela Mirro, Marilyn's daughter, painted the superb watercolors of the *Vanda* and *Ascocentrum* species. They add great aesthetic appeal and, in addition, give a better rendering of the details of the species than one ordinarily can get from a photograph.

Passing to those whose contributions are more directly embodied in the text, one person stands out above all others. To him, I owe an enormous debt of gratitude. He is Eric A. Christenson, who is without peer as an authority on the Sarcanthinae (syn.: Aeridinae) subtribe of the Vandeae tribe, which includes all the orchid species covered in this book. The main taxonomic portions would have been impossible without his input and cooperation. But, beyond that, his knowledge of natural habitats and cultural requirements is formidable. He also is a superb editor. His ready willingness to help in all these directions made the preparation of the manuscript a pleasure rather than a constant nightmare about accuracy and completeness.

Robert J. Griesbach, a geneticist in the field of horticultural research at the U.S. Department of Agriculture, and also chairman of the Research Committee of the American Orchid Society, guided me through the intricacies of the treatment of chromosomes and genes, and their relevance to the rest of the book. He kindly reviewed an early draft of the chapter on the subject and corrected the initial errors and confusions—all with his characteristic good humor and precision.

Robert H. Hesse, affiliated with the Research Institute for Medicine and Chemistry, located in Cambridge, Massachusetts, and an accredited judge of the American Orchid Society, was my consultant on matters involving an expert knowledge of chemistry. Like the others, he always was generous with his knowledge and its application to the cultivation of orchids.

Carl Withner, an eminent botanist and authority on orchids, reviewed the Guide to Orchid Classification and Nomenclature, which appears as Appendix A, and he contributed greatly to Chapter 2, on chromosomes and genes. He also meticulously reviewed an earlier draft of the entire text.

Albert Rutel, whose avocation is helping others grow orchids well, voluntarily sought out and uncovered a wealth of useful material about Malayan orchids that

I probably would not have discovered on my own. Roy Tokunaga and Douglas Schafer of Honolulu kindly supplied some Hawaiian material of considerable historical significance.

My thanks also to Richard E. Kaufman, who read an early draft of the book and made helpful comments, mainly about style but sometimes about substance as well.

Of course, opinions expressed in this book, along with any errors of commission or omission, are solely mine, and responsibility should in no way be attributed to the individuals who helped me.

Finally, a salute to my wife, Lois, who for more than two years grudgingly tolerated the disorderliness of several rooms in our house that were strewn with books and papers related to the project. My feeling of relief on finally completing the manuscript pales in comparison with hers.

# PART ONE

# The Species and Their Hybrids

# 1

# The *Vanda* Species

## ORIGIN AND COMPOSITION OF THE GENUS

*Vanda* is a genus in the subtribe Sarcanthinae (syn.: Aeridinae) of the Vandeae tribe. It is closely related to the genera *Aerides, Ascocentrum, Holcoglossom, Papilionanthe,* and *Rhynchostylis,* and to certain other Asian monopodial orchids.

The word *Vanda* is derived from ancient Sanskrit and originally referred to the sacred mistletoe found on oak trees, the oak being *Vandaca.* The term was adopted by Sir William Jones in 1795 to cover epiphytes in general, including orchids, in his observations about the flora of northeast India, published in *Asiatic Researches* (1795, 4:302). The name was retained by Robert Brown when he established the genus *Vanda* in 1820, based on the orchid that today is known as *Vanda tessellata,* but which originally was known as *Vanda roxburghii.* (Today, the two are synonymous.) Reflecting its history, the full botanical name of the genus is *Vanda* Jones ex R. Brown.

*Vanda* species are found only in the Eastern Hemisphere. The range of distribution within that part of the world spreads in a broad arc from northern India to the Mariana Islands and southward to southern India, Sri Lanka, and even to the Northern Territory of Australia. Within that vast area of many macro- and microclimates, the locales of the individual species usually are quite restricted, as they are for other orchid genera, such as *Phalaenopsis.* As a general rule, the wider the geographical range of a particular species, the more variable are its floral characteristics from place to place. This is especially true of flower color and patterns. The variability creates problems both for taxonomists and for those who wish to study hybridization records. In some cases, the provenance of a given plant can play a critical role in its use as a parent.

One might surmise that it should be easy to compile a list of the currently recognized *Vanda* and *Ascocentrum* species. That is not so. The taxonomy of both

19

these genera is in a very unsatisfactory state. Two principal reasons account for the sad state of affairs, both of which augur badly for the future. First, there has been far too little collecting of scientific material, a situation that can only become much worse, given the overwhelming rate of destruction of habitats. Second, too few botanists are qualified to undertake the needed explorations and studies, and the very few who are both qualified and interested receive inadequate support. The current situation deserves urgent attention, because soon it will be too late. Although some species indeed have been collected in quantity and introduced to horticulture, very few specimens have been preserved with adequate information about their precise locations in the wild and about the range of the species. For example, some of the species we regard as obscure have been seen only once or a few times. In these circumstances, no conclusions can safely be drawn about the variability and boundaries of the species.

Among the unsettled issues is whether we should treat the species *Vanda sanderiana* as a *Vanda* or, as some botanists do, as a species belonging in a separate genus of its own, *Euanthe*. This book adopts the first of these two alternatives, primarily for reasons of convenience. In any event, it is accepted practice in nearly all *horticultural* literature to call it a *Vanda* species.

Another area of differing treatment pertains to the so-called terete species and semi-terete hybrids. Nearly all botanists place the terete-leaf species in *Holcoglossom* or *Papilionanthe*. In much of the horticultural literature, however, they still are treated as "vandas"; the other vandas are called "strap-leaf" vandas. The treatment of vandas in this book is limited to strap-leaf plants, for several reasons. First, the vast majority of *Vanda* species and hybrids in cultivation outside the Southeast Asian countries fall in this group, partly because they bloom well with less heat and less intense light than are required by the terete forms. Second, except for the cut-flower trade of Southeast Asia, the main interest of *Vanda* hybridizers and growers since the 1950s has been directed toward perfecting strap-leaf vandas, and not toward improvement of the terete and semi-terete hybrids. Finally, and, in some respects, most importantly, the exclusion of these from the genus *Vanda* is in keeping with the conclusions of the leading taxonomic authorities in the field (Garay 1972, 1974; Christenson 1986a, 1987; and Seidenfaden 1988). All of them have placed the terete species in *Holcoglossum* and *Papilionanthe*.

An added complication is that taxonomic revisions becloud the historical record—not so much disputed revisions but ones that are almost universally accepted. For example, *Vanda suavis* was a widely used parent of many of the early hybrid vandas, but subsequently it was redefined as being synonymous with the species *V. tricolor* (although it generally is given varietal status). Thus, the hybrid

*V.* Burgeffii (*V. sanderiana* × *V. suavis*), registered in 1928, today would be regarded as the same as the hybrid *V.* Tatzeri (*V. sanderiana* × *V. tricolor*), which already had been registered in 1919. Similarly, *V. roxburghii*, the type-species of the genus, now is regarded as being the same as *V. tessellata* (and without varietal status), yet many hybrids are on the record as having *V. roxburghii* as a parent. (See Appendix A for an explanation of varietal status.)

Given the differences among taxonomists, along with fully accepted revisions, preparing a list of the *Vanda* species required the assistance of a specialist. Eric A. Christenson, a much-cited taxonomic authority on the *Aerides-Vanda* alliance, graciously came to the rescue by providing his personal list, which he later presented at the Fourteenth World Orchid Conference, in Glasgow, in April 1993. The list is reproduced here as Appendix B—Christenson's list of the genus *Vanda*. His synopsis of *Vanda* Jones ex R. Brown includes *Euanthe* as a subgenus of *Vanda*. It also unites *Trudelia* with *Vanda*; he places *Trudelia* in the *Vanda* section *Cristatae* (Lindl.), which includes *Vanda alpina, V. chlorosantha, V. cristata, V. griffithii, V. javieriae,* and *V. pumila.*

One measure of horticultural interest in a species, and of its availability, is whether it ever has received an award of any sort from the American Orchid Society. One of the functions of the Society's formal judging system is to "recognize" and record orchid species regarded by the judges as having botanical or horticultural interest, provided they are presented for such consideration at one of the Society's judging sessions. Not all the *Vanda* species on Christenson's list of the genus *Vanda* have received an award of recognition from the American Orchid Society (AOS). I have flagged the species that had been so recognized by the end of 1994, or that had received some other AOS award for flower quality or culture, which implies recognition. Those species not yet recognized by the AOS can safely be presumed to be rare or even nonexistent in cultivation, at least in the United States.

Another test of horticultural importance is to enquire which of the *Vanda* species have been used, at some time or other, as a parent in the making of a hybrid. Given more than half a century of active *Vanda* hybridizing and the inquisitive nature of orchid enthusiasts, one would expect that very few available *Vanda* species would remain ignored. The *Vanda* species that have been used as a parent at least once, and the resulting hybrid recorded, also have been marked in Appendix B.

If we exclude *Vanda spathulata*, which Christenson (1992, 90–91) has transferred to a new genus, *Taprobanea*, we find that, by the end of 1994, 27 *Vanda* species on Christenson's list had caught the eye and the toothpick of at least one eager breeder and had been crossed with some other strap-leaf *Vanda*, and registered. Those species comprise nearly all the *Vanda* species recognized by the American Orchid Society, and, in addition, include some that apparently are sufficiently uncommon that they have never been presented for recognition at the Society's formal judging sessions, or at least have never been awarded.

## DESCRIPTIONS OF *VANDA* SPECIES

Descriptions of the *Vanda* species are scarce and sometimes contradictory in the literature. The color and markings of the flowers of some of the *Vanda* species are variable within the species, especially from one locale to another. This variability makes for a rich pigmentation pool on which hybridizers may draw, but it complicates horticultural records and descriptions. Moreover, while a botanist may decide that certain plants belong together in the same species, the flowers may look quite dissimilar in some respects to those of us whose interest is purely horticultural. For example, flower color and markings are especially important to horticulturists, but to taxonomists they are of much less significance than are some of the morphological details of the flower that go little noticed by most orchid growers.

One cannot always rely on the so-called type-description written by the botanist credited with naming that species. Unfortunately, type-descriptions sometimes are not typical of the plants that most growers will encounter. As Robert L. Dressler (1981, 276) observes, type-specimens "are of historical value and serve as landmarks to determine the application of names. The first specimen to be collected and named may be a very unusual one, but this does not affect its status as a type-specimen."

The Royal Horticultural Society in London does not provide any descriptive information about the species parents of the hybrids it registers, nor does it record the form or variety of a species listed as a parent, if in fact a form or variety was involved. Nor is there any other recognized authority on forms and varieties; certain such names are well known and widely recognized (for example, the *alba* form of *Vanda sanderiana,* which usually is erroneously called "var. *alba*"), while others are not widely or uniformly accepted.

Another consideration to bear in mind is that plants submitted for judging or exhibited in orchid shows usually are presented because their owners consider them to be different in some favorable respect; in other words, better than what is typical. The plant may be a polyploid (i.e., have more than the normal number of chromosomes), or an unusual color form, or simply a superior example of the species because of a fortunate combination of genes. If so, it may have more or larger flowers, of better shape, color, arrangement, and substance than is typical of the species. Similarly, hybridizers try to make progress by using breeding stock that is exceptionally good in as many respects as possible, rather than by using ordinary specimens of species or hybrids. As a general proposition, the more a species is given to great variability in its characteristics, the more hybridizers are likely to experiment with its superior forms, because the variability connotes a plasticity that may produce something interesting in the hybrid offspring.

The following descriptions of the better-known *Vanda* species describe typical plants of the species, not the exceptional clones. Separate forms and varieties are mentioned only when they have received a reasonable amount of attention in the literature. Many authors confuse "varieties" with "forms," and in a number of cases it is not clear which is correct. The descriptions were compiled from many sources, some of them confusing in their details and occasionally at odds with one another about such matters as number and size of flowers, length of the inflorescence, and time of blooming. In some instances, information in the *Awards Quarterly* of the American Orchid Society was used for reference purposes, even though that information refers to awarded, and therefore generally superior, plants. What is presented, in short, is essentially a set of composite descriptions drawn from more than one source, and supplemented by my personal exposure to some of the species.

## *Vanda alpina* (Lindl.) Lindl.

GENERAL LOCALE: The Himalayas from Nepal to Bhutan and the Khasia Hills of India, mostly at elevations of 900–1200 m (3000–4000 ft), where, in winter, hoarfrost and occasionally even snow occur.

FLOWERING TIME: Spring, summer.

DESCRIPTION: Dwarfish, densely leafy plant about 18 cm (7 in) tall; the leaves are proportionately short—10–13 cm (4–5 in). Wrongly described in the first edition of Bechtel, Cribb & Launert (1981, 410) as being terete (but corrected in the third edition, 1992). Very short inflorescences with 1–4 faintly fragrant, waxy flowers of about 4 cm (1.5 in) vertically. The petals are narrow and inward-curving. Sepals and petals are light yellow-green. The lip is fleshy. The side lobes are erect and triangular; they are purplish black inside and pale yellow or greenish yellow outside. The midlobe of the lip is yellow with blackish purple stripes. *Vanda alpina* resembles *V. cristata.*

This species has, on occasion, been treated as *Trudelia alpina* (Lindl.) Garay (1986), but Christenson (1992) places it in *Vanda,* in the *Cristatae* section.

## *Vanda bensonii* Batem.

GENERAL LOCALE: Burma and Thailand. Veitch (1887–1894, 2:89–90) states that, in lower Burma where it was first found, it grows "on trees in a deciduous jungle fully exposed to the sun in the dry season, when the temperature frequently rises to 45°C (112°F) in the shade and when its leaves are often scorched."

FLOWERING TIME: Mainly spring, but also reported as flowering at other times.

DESCRIPTION: A small plant with erect 30 cm (12 in) stalks. Leaves are 18–25 cm (7–10 in) long. Flowers are about 5 cm (2 in) across, produced on 30–36 cm (12–14 in) inflorescences bearing 10–20 scented flowers. The petals are twisted and recurved; the sepals also are recurved. The sepals and petals are greenish yellow with chestnut-brown reticulation; their reverse side is pale pink or white. The lip is light purple with darker striation at the base of the midlobe; the side lobes are white.

*Vanda bensonii* occasionally forms natural hybrids with *V. coerulea*. These are named *V.* × *charlesworthii*.

## *Vanda brunnea* Reichb. *f.*

GENERAL LOCALE: Burma, northern Thailand, and possibly China.

FLOWERING TIME: Autumn, but variable under cultivation.

DESCRIPTION: A variable species about which there is considerable ambiguity with respect to its relationship to *Vanda denisoniana* var. *hebraica*. The leaves are closely spaced and about 18 cm (7 in) long and 2.5 cm (1 in) wide. The erect inflorescence is about 50 cm (20 in) in height or longer, often with 15 or more widely spaced flowers that are about 3.2 cm (1.25 in) across and have very long, thin pedicels. The sepals and petals are rather wavy edged and reflexed, greenish in color, with a brown overlay and maroon-brown spots, tessellations, or bars. The lip is light olive-green, bifurcated and prominent. The flowers have a strong fragrance.

A description of one plant awarded a Certificate of Horticultural Merit by the American Orchid Society, blooming in March, described the lip as "narrow, purple, with a brown, flaring, forked tip."

## *Vanda coerulea* Griff. ex. Lindl.

GENERAL LOCALE: northeast India (Khasia Hills, Assam), northern Burma, northwest Thailand, and China (Yunnan), at exposed locations at 900–1500 m (3000–5000 ft) elevation, on small deciduous trees.

FLOWERING TIME: Mainly autumn, but in cultivation may bloom several times yearly, and in any month.

DESCRIPTION: A strong-growing plant with stout vegetative stems that can reach 1.5 m (5 ft) in height, but plants may bloom when less than 30 cm (12 in) tall. The leaves are 15–25 cm (6–10 in) long and issue from the plant's stem with a more open angle at their basal portion than do those of other vandas. The flowers typically are 9–10 cm (3.5–4 in) across and of moderate substance. The inflorescence is 40–75 cm (15–30 in) long and normally bears 10–14 flowers held well above the leaves and well spaced on the flower stalk. The background colors of the sepals and petals range from nearly white to pale violet, with tessellation that can vary from very faint blue to deep violet (see Plates 1, 13, and 14). There is a rare all-pink form with a vivid deep pink lip (Plate 15), as well as an exceedingly rare *alba* form (Plate 16) with a snow-white lip.

The petals characteristically twist about 90 degrees on their basal axis, although the petals of some plants of *Vanda coerulea* show much less tendency to twist than do others. Most plants have sepals and petals that are considerably longer than their width, but some superior plants have very round sepals and petals. The individual flowers are much paler and smaller when they first open than when they are mature. If a flower loses its pollinia, it will turn white within a day or two. (Pollinia are cohesive masses of pollen grains, located in the tip of the column of the flower.)

The lip of *Vanda coerulea* is dark purple. Each side lobe terminates in a needle-sharp, backward-curving "hook" (see Plate 17). This feature is very useful for distinguishing a flower of the pure species from that of a closely related hybrid; the hook becomes shorter and blunter, and may even not be present at all, in hybrids that in other respects resemble *V. coerulea* (see Plates 18 and 19).

*Vanda coerulea* is known to produce two natural hybrids: *Vanda × amoena*, with *V. tessellata*; and *Vanda × charlesworthii*, with *V. bensonii*. (In the name of a natural hybrid within a genus, a multiplication sign precedes the second term of the name, which is written like that of a species—see Appendix A.)

*Vanda coerulea* is native to higher altitudes and colder habitats than most *Vanda* species can tolerate. The plants cannot survive very long in the hot, humid conditions that prevail throughout the year in places like Bangkok and Singapore. They prefer daytime humidity in the neighborhood of 60%, whereas most other vandas grow better with humidity close to 80%. In their native habitat in winter, frost is not uncommon. The days can be very bright and dry. For greater detail about the climate of the native habitats of *Vanda coerulea*, see Pradhan (1973): *The Habitat and Growing Conditions of Vanda Coerulea in Nature, with Cultural Hints Gleaned Therefrom*, and Suprasert (1975): *Notes on Vanda Coerulea in Thailand*.

The species is easy to grow and to flower in temperate climates under conditions similar to those provided for cattleyas. The line-bred plants presently available are far superior to collected plants and should be more widely grown, for no other *Vanda* species can approach *V. coerulea* in beauty, floriferousness, and ease of cultivation in temperate climates.

### *Vanda coerulescens* Griff.

GENERAL LOCALE: Northeast India, Burma, Thailand, China (Yunnan), at 300–400 m (1000–1500 ft) elevation.

FLOWERING TIME: Spring, early summer.

DESCRIPTION: A compact plant, 0.3–0.6 m (1–2 ft) tall, with leaves 13–18 cm (5–7 in) long. Flowers are 2.5–3.8 cm (1–1.5 in) across, on almost upright inflorescences bearing many flowers. The pedicels are about 3.8 cm (1.5 in) long. The sepals and petals are very pale bluish violet; the petals twist backwards, like those of *Vanda coerulea*. The small lip has a darker blue-violet midlobe. The side lobes are light purple.

### *Vanda concolor* Blume

GENERAL LOCALE: China.

FLOWERING TIME: Winter, spring.

DESCRIPTION: An erect plant, 0.3–0.9 m (1–3 ft) tall, with leaves up to 25 cm (10 in) long and 2.5 cm (1 in) wide. Arching inflorescences bear 7–10 flowers measuring about 5 cm (2 in) across. They are greenish brown, and their sepals and petals are wavy. The lip is yellowish, with rosy red dots or streaks on the side lobes. The flowers are fragrant and long lasting. The species is exceedingly rare in cultivation.

### *Vanda cristata* Lindl.

GENERAL LOCALE: Himalayas from Nepal to Bhutan, north to Tibet, and south to Khasia and Bangladesh, at rather high elevations.

FLOWERING TIME: Spring, summer.

DESCRIPTION: A short, stout plant, about 20 cm (8 in) tall. The inflorescence also is short, with 4–6 waxy, fragrant flowers measuring 4–5 cm (1.5–2 in) across, on long pedicels. The sepals and petals are uniform yellow-green. The prominent lip, which is its most striking characteristic, is yellow or creamy with deep maroon or dark purple stripes (see Plate 20). The sepals and petals are narrow and incurved. *Vanda cristata* is closely related to *V. alpina* and *V. pumila*. See the reference to *Trudelia* in the discussion of Christenson's list earlier in this chapter.

### *Vanda dearei* Reichb. f.

GENERAL LOCALE: Borneo, in the lowlands.

FLOWERING TIME: Mostly summer and early autumn.

DESCRIPTION: A rather large plant, with 3–6 fragrant flowers on a short inflorescence (see Plate 2). The flowers are about 4.5 cm (1.75 in) across; the sepals and petals are about half as wide as they are long. The flowers usually have a cream-colored background overlaid with a dull greenish brown, but on some plants they are yellow, like *Vanda denisoniana*. The sepals and petals fade to creamy white basally. The base of the lip, and the side lobes, also are white; the midlobe is yellowish. Most have somewhat bent-back sepals and petals, while a few have relatively flat ones, which makes for pleasing, fairly full-shaped flowers. *Vanda dearei* tends to transmit its rich, sweet fragrance to its immediate progeny. It is perhaps the most cold-sensitive species of *Vanda,* and some yellow-flowering hybrids with *V. dearei* ancestry share this cold-sensitivity.

Considering the influence *Vanda dearei* has had in *Vanda* hybridizing, it is somewhat remarkable that it is not better known and more fully described in the botanical and horticultural literature. The explanation probably is that it is difficult to bloom in temperate climates because of the need for elevated temperatures with little fluctuation.

### *Vanda denisoniana* Benson & Reichb. *f.*

GENERAL LOCALE: Burma, China (Yunnan), northern and northeastern Thailand, at 600–750 m (2000–2500 ft) elevation, where there is heavy annual rainfall.
FLOWERING TIME: Mostly spring.
DESCRIPTION: A medium-size plant with a twisting growth habit. The inflorescences are about 15–18 cm (6–7 in) long, arching or horizontal in habit, with up to 8 waxy, long-lasting flowers measuring about 5–6.5 cm (2–2.5 in) vertically (see Plate 3). The flowers are delightfully fragrant at dusk. The color of the sepals and petals ranges from pale creamy yellow to greenish yellow. The lip is broad and yellow-green, with rounded side lobes. Kamemoto and Sagarik (1975, 167) show color photos of six color forms of the species, including one with light copper-brown sepals and petals. The shape of the flower also varies considerably. Some varieties have round and quite flat petals while others, especially the darker colored varieties, have elongated, recurved petals and sepals. Some varieties have spots, short bars, or tessellation on the petals and sepals (e.g., the variety *hebraica*). There is dispute whether these are properly classified under *Vanda denisoniana* or whether they should be included under *V. brunnea* or elsewhere.

### *Vanda foetida* J. J. Smith

GENERAL LOCALE: Sumatra.
FLOWERING TIME: Mostly summer.
DESCRIPTION: Resembles *Vanda dearei* in habit. It has a short inflorescence, with 2–3 flowers measuring about 4.5 cm (1.75 in) across. The flowers are waxy. The rather broad sepals and petals usually are mauve at their edges, shading to cream color at their base, and often are somewhat reticulated with darker mauve. The lip has yellowish side lobes and a midlobe that is bilobed and more or less purple or mauve-purple in color. The flowers have been described as having a strong, unpleasant odor; hence the name.

### *Vanda hastifera* Reichb. *f.*

GENERAL LOCALE: Borneo, in full sun.
FLOWERING TIME: Almost continuously.
DESCRIPTION: The plant has a short inflorescence with 3–8 long-lasting, fragrant

flowers normally measuring about 5 cm (2 in) across. The flowers are pale yellow, heavily blotched with dark brown. The plant has a scrambling growth habit.

A plant receiving a Certificate of Botanical Recognition from the American Orchid Society had 5 flowers on one inflorescence. It was described in the *Awards Quarterly* (20[4]:176) as follows:

> Flowers pale yellow with an overlay of brown tessellations and blotches, with a distinct yellow margin. Lip with white side lobes and magenta marking on interior midlobe, with one reflexed basal tooth and central portion fleshy-greenish; dorsal surface hairy. Natural spread of flower 5.4 cm (2.13 in).

The same issue of the *Awards Quarterly* (20[4]:199) described another *Vanda hastifera* that received a Certificate of Horticultural Merit. It had 8 flowers, slightly larger (5.9 cm [2.3 in] vs. 5.4 cm [2.13 in]), of

> light greenish yellow with amber spots along midline of sepals and petals; spotting and slight tessellation of same hue becoming darker at base of lateral sepals; sepals and petals clawed, with wavy margins; lip narrow, midlobe green below, side lobes white with inside deep rose extending down midline and terminating at two globules on apex; substance heavy.

*Vanda hastifera* is very closely related to *V. celebica* Rolfe, *V. lindeni* Reichb. *f.*, and *V. scandens* Holtt.

## *Vanda insignis* Blume

GENERAL LOCALE: Moluccas, Timor, Malayan Archipelago, Java, at low elevations, on small trees with little shade.
FLOWERING TIME: Variable.
DESCRIPTION: The inflorescences bear 4–7 flowers, measuring 5–6.5 cm (2–2.5 in) across. The sepals and petals are much recurved. Their color is tawny yellow with brown blotches that coalesce toward the sepal and petal edges. The lip is one of the largest in the genus. It has two white side lobes; the midlobe is mauve or rose-purple, broad and almost flat, though somewhat upcurved near the rounded tip, giving it the appearance of the blade of a shovel. According to Holttum (1972, 724), it flowers well in both the mountains and lowlands of Java, but not so well in the uniform climate of Singapore.

## *Vanda lamellata* Lindl.

GENERAL LOCALE: The Philippines, widely distributed in hot, humid locales; also found in the Mariana and Ryuku Islands, Taiwan, and Sabah, Borneo.
FLOWERING TIME: Variable; the plants can flower during most months of the year, and several times.

DESCRIPTION: Plants grow to about 40 cm (16 in) tall. Their erect inflorescences, about 25–35 cm (10–14 in) long, bear 8–15 flowers measuring about 4 cm (1.5 in) vertically, on long, thin pedicels. The sepals and petals are recurved and backward-tilting. They have a light yellow or greenish background, with chestnut-brown spots or stripes. The lip is short and fleshy. The flowers are faintly fragrant, especially in the morning.

Given its wide geographical distribution, it is not surprising that there are several varietal forms. The most attractive is the variety *boxallii* (see Plates 4 and 21). It generally has more and larger flowers, measuring 4–5 cm (1.5–2 in) vertically, brightly colored, arranged cylindrically on 30–46 cm (12–18 in) flowering stalks. The background color of the flowers is creamy white or light yellow; the lateral sepals, on their outer side, are the same color as the petals and the dorsal sepal, while their inner half is purplish and crimson. The lip is rose-purple. This is the variety most often used in hybridization.

Another variety, *remediosae,* is a native of Jolo and Mindanao in the Philippines. It blooms three times a year. Its petals and dorsal sepals are white; the lateral sepals are yellow, and their inner half is citron-green with buff-citrine stripes.

Still another variety, *calayana,* has even more flowers, on a long raceme.

Valmayor (1984, 2:220–233) has some excellent full-page color photos of the inflorescences of several varieties of *Vanda lamellata.*

## *Vanda laotica* Guill. = *V. lilacina* Teijsm. & Binn.

Considered by Seidenfaden and Christenson to be the same as *Vanda lilacina;* listed here only because it appears as a parent in *Sander's List of Orchid Hybrids.* See *V. lilacina.*

## *Vanda lilacina* Teijsm. & Binn. (Syn: *V. laotica*)

GENERAL LOCALE: Burma, Laos, Cambodia, Vietnam, and China (Yunnan).
FLOWERING TIME: Winter or spring.
DESCRIPTION: About 8 or more small flowers measuring nearly 2.5 cm (1 in) across, on a rather long raceme; the petals are twisted about 90° on their axis and are somewhat recurved. The sepals and petals are white; the lateral sepals sometimes are purple-tipped. The side lobes of the lip have purple dots inside; their front edge is white. *Vanda lilacina* is an interesting parent; its hybrids can produce several inflorescences, even on young plants.

## *Vanda limbata* Blume

GENERAL LOCALE: Java and in the foothills of central Luzon in the Philippines, in areas of high temperature and humidity.
FLOWERING TIME: In Java, July–August. In the Philippines, in June.

DESCRIPTION: A robust plant infrequently seen in cultivation. The inflorescence is 15–25 cm (6–10 in) long and bears 10–12 shiny, scented flowers of about 4–5 cm (1.5–2 in) across. The sepals and petals are nearly equal in size and are similarly shaped; their color is yellowish with tessellations and flecks of light cinnamon-brown or chestnut, which suffuse into adjacent spaces. The flowers generally have a narrow yellow border around their edges. The lip is rose-lilac, with a white border.

Note that an unrelated orange-brown-flowered species of unknown origin recently has been exported from Southeast Asia under the name *Vanda limbata*.

Refer to Comber (1982a) for photos of *Vanda limbata* in its native habitat.

## *Vanda liouvillei* Finet

GENERAL LOCALE: Burma, northern Thailand, and Laos.
FLOWERING TIME: Spring.
DESCRIPTION: The inflorescence is up to 50 cm (20 in) long, with approximately a dozen or more widely separated flowers measuring about 4 cm (1.5 in) across, on long pedicels of about 5.5 cm (2.25 in). The sepals and petals are undulated, recurved, and twisted. There is much variability in color, but most commonly the flowers have yellowish sepals and petals with fine brown dots or spots arranged in parallel rows starting from the basal portion. The lip is fish-tailed; the side lobes usually are white with yellow tips, minutely speckled with pale purple; the midlobe is red-purple.

## *Vanda luzonica* Loher ex Rolfe

GENERAL LOCALE: The Philippines, on Mt. Pinatubo in Zambales Province, and also some other localities, at medium elevations.
FLOWERING TIME: In the first and fourth quarter of the year, but variable under cultivation elsewhere. (In Florida, however, it is very seasonal.)
DESCRIPTION: A leggy plant of up to 1.2 m (4 ft) in height; the plant stem usually is twisting or bent; the leaves are about 40 cm (16 in) long and 2.5 cm (1 in) wide. The inflorescence is about 40 cm (16 in) long and bears 10–25 flowers measuring 5–7.5 cm (2–3 in) across, on rather long pedicels. The petals are twisted. Both sepals and petals are recurved, waxy, pure white or creamy white, with irregular blotches of rose-lilac near the tips and a thin ring of magenta at the base, close to the column. The lip is magenta-purple, with even darker longitudinal lines. (See Plate 5.) *Vanda luzonica* appears to be closely related to *V. tricolor,* and the two sometimes are confused, given the variability of *V. tricolor;* however, while *V. luzonica* has no fragrance, *V. tricolor* is very fragrant. *Vanda luzonica* is one of the most attractive *Vanda* species.

## *Vanda merrillii* Ames & Quisumbing

GENERAL LOCALE: The Philippines, in Quezon Province on the island of Luzon, at medium elevations.

FLOWERING TIME: Two or three times a year.

DESCRIPTION: A tall, leggy plant. It often issues several racemes of about 25 cm (10 in) long, bearing 9–16 fleshy, lacquered-looking, fragrant, strikingly attractive flowers measuring about 4–5 cm (1.5–2 in) across (see Plate 6). The sepals and petals are narrowed at their base and have a yellow background heavily overlaid with carmine-red stains and blotches, especially at the tips. The lip is yellow with a red spur; the midlobe is splashed with red. The flowers have very long pedicels.

At least two varieties exist. The variety *immaculata* has yellow flowers that are slightly larger than the standard type and lack pronounced bars or blotches. The variety *rotori* also lacks bars. Its sepals and petals are oxblood-red; it has a red midlobe.

## *Vanda parviflora* Lindl. = *V. testacea* (Lindl.) Reichb. *f.*

Considered by Seidenfaden and Christenson to be the same as *Vanda testacea*; listed here only because it appears as a parent in *Sander's List of Orchid Hybrids*. See *V. testacea* and Plate 22.

## *Vanda pumila* J. D. Hook.

GENERAL LOCALE: Northwest Himalayas, Nepal, Sikkim, northeast India, Burma, northern Thailand, Laos, Vietnam, China (Yunnan), and in the uplands of northern Sumatra, at elevations of 1100–1300 m (3500–4300 ft), under conditions of bright light. In the Sikkim Himalayas it has been collected in hot valleys at 600 m (2000 ft) altitude.

FLOWERING TIME: Spring, but can flower in any month.

DESCRIPTION: The heavily scented flowers measure nearly 5 cm (2 in) vertically and somewhat less horizontally. Typically, this species has 2–4 flowers borne on a short, almost horizontal inflorescence. The petals and sepals have wine-purple stippling at their base. The lateral sepals curve forward in a pronounced fashion; the petals twist to a horizontal position. The lip is concave and has longitudinal purple or maroon stripes or stippling. The lip is large in relation to the flower size and is the most attractive feature of this species. *Vanda pumila* cannot tolerate lowland tropical conditions, according to Comber (1982b).

## *Vanda roeblingiana* Rolfe

GENERAL LOCALE: The Philippines, in Mountain Province on the island of Luzon, at about 1200–1500 m (4000–5000 ft) elevation. Also found on the Malay Archipelago. Cool growing.

FLOWERING TIME: Usually summer.

DESCRIPTION: The plant grows to 50–70 cm (20–28 in); it has a 18–30 cm (7–12 in) inflorescence that typically bears 8–15 flowers measuring about 5 cm (2 in) across. The sepals and petals are yellow, heavily overlaid with reddish brown blotches and stripes (see Plate 7). The lip is relatively large; it has a bifurcated mid-lobe that is yellowish with red stripes and is covered with fine hairs. The midlobe has fringed margins. Like other montane Philippine orchid species, such as *Phalaenopsis lindenii, Vanda roeblingiana* requires a cooler climate than do the lowland Philippine vandas.

## *Vanda sanderiana* Reichb. *f.*

GENERAL LOCALE: Philippines, on the island of Mindanao, primarily in the Province of Davao, at low elevations of quite constant high heat and humidity.

FLOWERING TIME: Usually in autumn, although in cultivation it may occasionally bloom at some other time.

DESCRIPTION: Tall-growing, erect plant with closely spaced, wide, stiff leaves of 30–38 cm (12–15 in) in length. The racemes also are erect and stout, and ordinarily bear 7–10 flowers, 9–10 cm (3.5–4 in) across. The flowers of *Vanda sanderiana* differ from those of other *Vanda* species in several respects (see Frontispiece and Plate 23). They are flat, and the petals and sepals are wider in relation to their length (the flowers of some *V. coerulea* plants are another exception in this latter respect). The underside of the basal portion of the lip of *V. sanderiana* is balloon-shaped, with no spur whatsoever—a critical distinguishing feature. (See Plates 24 and 25). The petals and dorsal sepal are pale lilac or pinkish white, with some cinnamon-colored speckling on their inner halves. The lateral sepals are larger than the dorsal sepal and are tawny yellow, with prominent veining of dark cinnamon or reddish brown. The lip is broad and usually is a dull tawny color streaked with brownish red.

There is a so-called *alba* form that has white petals and dorsal sepal, often with some light apple-green speckling basally. The lateral sepals are apple-green, with faint reticulation (see Plate 26).

As with *Vanda coerulea,* selective artificial propagation of *V. sanderiana* has produced vastly superior forms. A fine example is pictured in Plate 23. *Vanda sanderiana* provides the foundation on which most of today's fine *Vanda* hybrids have been developed.

Some taxonomists place *Vanda sanderiana* in a monotypic genus, *Euanthe,* mainly because of the complete absence of a spur on the underside of the basal portion of the lip; however, in horticulture, it is almost universally included in the genus *Vanda.* Christenson treats it as a monotypic subgenus of *Vanda.*

Because *Vanda sanderiana* appears in the ancestry of nearly all *Vanda* and most *Ascocenda* hybrids, and often more than once, the climatic conditions in the native habitat of this species are of special interest. The following description is adapted from an article by Robert Warne (1954); Warne lived on Mindanao for several years and collected many *V. sanderiana* for shipment to Hawaii. He was one of the leading pioneer breeders of vandas in Hawaii.

Warne reports that most of the collected plants of *Vanda sanderiana* came from near Davao, the principal city of the island of Mindanao, at elevations between sea level and 90 m (300 feet), although some plants at times were seen as high as 600 m (2000 ft). The plants grew near the tops of tall hardwood trees. The latitude of Davao is about 5° north of the equator, so there is very little difference between winter and summer temperatures. The daily temperature averages about 27°C (81°F). The daily high is about 31°C (88°F) and the low is approximately 23°C (73°F), practically year-round. Substantial rainfall occurs throughout the year— there is no "dry season." Typically, the sun shines in the morning, and rain falls in the afternoon. Relative humidity is almost constantly in the 80% to 90% range. (Table 1-1 presents monthly data.) Warne goes on to say that the plants

> were probably shrouded by clouds most of the afternoon, had broken light through the tree branches above them about half of the remaining time and were exposed to the full sun about one-fourth of the day for short periods of time.

Robert Warne's succinct description of the climate of the native habitat of *Vanda sanderiana* is well worth bearing in mind when growing today's *Vanda* hybrids, since nearly all of them have a strong representation of *V. sanderiana* in their

Table 1-1. Climate and weather of Davao, Mindanao (1903–1918).

| MONTH | MONTHLY RAINFALL | | MEAN DAILY TEMPERATURE | | MEAN DAILY RELATIVE HUMIDITY | MEAN DAILY CLOUD COVER |
|---|---|---|---|---|---|---|
| January | 13.7 cm | (5.4 in) | 26.1°C | (79°F) | 88% | 73% |
| February | 15.5 | (6.1) | 26.1 | (79) | 86 | 71 |
| March | 19.3 | (7.6) | 26.7 | (80) | 86 | 63 |
| April | 19.3 | (7.6) | 27.8 | (82) | 86 | 60 |
| May | 18.0 | (7.1) | 27.8 | (82) | 85 | 58 |
| June | 18.3 | (7.2) | 26.7 | (80) | 84 | 64 |
| July | 21.8 | (8.6) | 26.7 | (80) | 81 | 70 |
| August | 22.1 | (8.7) | 26.7 | (80) | 79 | 66 |
| September | 22.9 | (9.0) | 26.7 | (80) | 81 | 71 |
| October | 29.5 | (11.6) | 26.7 | (80) | 84 | 68 |
| November | 19.1 | (7.5) | 26.7 | (80) | 87 | 68 |
| December | 22.1 | (8.7) | 26.1 | (79) | 88 | 74 |

genes. At the same time, most of them also have some representation of *V. coerulea* in their ancestry, and the cultural requirements of *V. coerulea* are different. Some hybrids grow and flower better under conditions closer to those in the habitat of one of the two species, while some others do better under the indigenous conditions of the other species. The best conditions for many hybrids will fall in between. Only experience with individual plants can disclose their respective optimum conditions. Nonetheless, the native conditions provide some parameters.

### *Vanda spathulata* (L.) Spreng. = *Taprobanea spathulata* (L.) E. A. Christ.

This species is excluded from the genus *Vanda* by Christenson (1992, 90–91), who places it in a new genus, *Taprobanea*. It is described here, and also is included in Tables 3-3 and 3-4, for historical and horticultural reasons, especially since it is listed as a *Vanda* parent in *Sander's List of Orchid Hybrids*.

GENERAL LOCALE: Sri Lanka and southern India.

FLOWERING TIME: Throughout the year.

DESCRIPTION: Holttum (1972, 726) says the species has a "habit more like *Arachnis* than *Vanda*, with climbing stems and short leaves." The inflorescence is stiff, and about 20–30 cm (8–12 in) long, with a "succession of many flowers, only 1–3 open together." The flowers are about 3–4 cm (1.25–1.5 in) wide, "clear yellow, with almost flat, round-ended sepals and petals."

One AOS-awarded plant was described as having unusually large flowers of

> clear bright yellow with faint darker flush at edges of segments; lip yellow, lined and suffused with raw sienna; substance very heavy, form full. Natural spread of flower 5 cm (2 in). (See Plate 27.)

The species is remarkably dominant as a parent of hybrids, not only with plants in the genus *Vanda* but also when used with members of other genera, such as *Aerides* and *Papilionanthe*. It is reputed to be hexaploid (i.e., to have six rather than the normal two sets of chromosomes).

### *Vanda stangeana* Reichb. *f.*

GENERAL LOCALE: Manipur area of India and other sections of the Assam Himalayas, at elevations of 400–500 m (1300–1700 ft).

FLOWERING TIME: July in its native habitat; variable and free-flowering in cultivation.

DESCRIPTION: A short plant, only 15–30 cm (6–12 in) tall, with 7.5–10 cm (3–4 in) leaves. The inflorescences are very short, with 1 or 2 nodding, fragrant flowers that open only partially and measure about 1.9–2.5 cm (0.75–1 in) vertically. The sepals and petals are greenish yellow, oblong, and blunt at their ends, and are tessellated with dark purplish brown. The sepals and petals are about 1.3 cm (0.5 in) long; the sepals are broader than the petals. The lip is white with some yellow and mauve spots.

## *Vanda sumatrana* Schltr.

GENERAL LOCALE: Sumatra.

FLOWERING TIME: Summer.

DESCRIPTION: *Vanda sumatrana,* in its habit and its short, few-flowered inflorescences, resembles *V. dearei,* and it has flowers of similar shape and size, about 4 cm (1.5 in). The sepals and petals are glossy and waxy; their color ranges from olive-brown at the tips to reddish brown at the base. The base of the lip is white, almost as in *V. dearei;* the tip of the lip is suffused with light brown. The flowers, though having a shiny texture, are a dull color. Their fragrance is not as good as that of *V. dearei,* but the lip may be regarded as better shaped. *Vanda sumatrana* flowers well in the hot, uniform climate of Singapore, where at least two varieties have been imported. The fragrance has been described as "creosote-scented." See Comber (1982b, 204–206) for some excellent color photographs of *V. sumatrana,* including one of a plant growing on rocks on the banks of Lake Toba in northern Sumatra, at about 850 m (2800 ft) elevation.

## *Vanda tessellata* (Roxb.) W. J. Hook. ex G. Don

GENERAL LOCALE: Northeast India, Nepal, Burma, southern India, and Sri Lanka.

FLOWERING TIME: Mainly summer. Some plants in cultivation bloom practically throughout the year.

DESCRIPTION: The plants usually reach 60 cm (2 ft) in height, but may be taller. The inflorescences are up to 30 cm (12 in) long, with 5–10 long-lasting, heavily perfumed flowers of about 5 cm (2 in) in width. The margins of the sepals and petals are undulated. There are many color forms, but most commonly the flowers have greenish yellow sepals and petals accompanied by dull brown tessellation. The lip has white side lobes with purple spots; the midlobe is dull violet and is paler at the base. (See Plate 8.) The arrangement of the flowers is better than that of a number of other *Vanda* species.

*Vanda roxburghii,* once considered a separate species and much used by early hybridizers, now is considered synonymous with *V. tessellata* (and without varietal status).

*Vanda tessellata,* then called and written as *V. roxburghii,* in honor of Dr. William Roxburgh, an early pioneer of Indian botany, was the species used in establishing the genus *Vanda.* It was the first *Vanda* cultivated in England; Sir Joseph Banks flowered it in the autumn of 1819.

*Vanda* × *amoena* is a natural hybrid of *V. tessellata* and *V. coerulea.*

### *Vanda testacea* (Lindl.) Reichb. *f.* (See *V. parviflora*)

GENERAL LOCALE: Widespread in India through Nepal and the Himalayas eastward to Manipur; Burma, Sri Lanka, and China (Yunnan).

FLOWERING TIME: Spring.

DESCRIPTION: The flowers of *Vanda testacea* are nearly 2.5 cm (1 in) across and have long pedicels. (See Plate 22.) The sepals and petals are twisted and recurved, and yellowish or pale orange in color. The lip has a dotted purplish violet midlobe with a yellow front edge. A considerable number of forms occur with respect to sepal and petal colors and in the shape of the lip. Color forms described by various writers range from brick-red through deep orange to white. A cultivar awarded a Certificate of Horticultural Merit by the American Orchid Society in 1985 (but labelled as *V. parviflora*) had 26 flowers and 4 buds on one inflorescence. A drawing shown by Seidenfaden (1988, 207), however, shows a short inflorescence with only about 6 flowers; this is in keeping with another plant (also awarded a Certificate of Horticultural Merit, but as *V. testacea*) which had 48 flowers and buds on seven inflorescences, making an average of 7 per inflorescence.

### *Vanda tricolor* Lindl.

GENERAL LOCALE: Indonesia, in east Java and Bali, at altitudes of 700–1700 m (2300–5500 ft), in the high branches of trees in semi-open situations; dubiously in Laos and Australia.

FLOWERING TIME: Anytime during the year, but normally from June to October, in the dry season.

DESCRIPTION: *Vanda tricolor* is a large plant with very attractive flowers and several varieties, the most influential of which, so far as hybridizing is concerned, is the variety *suavis,* which comes from east Java and Bali. In horticultural circles, the latter variety for many years was considered to be an independent species. (See Plate 9.)

The inflorescences on the variety *suavis* are 25 cm (10 in) or longer and are slightly arching. Ordinarily they bear 6–12 very fragrant flowers measuring about 5–8 cm (2–3 in) vertically and having good substance. The sepals and petals are rather narrow at their base, have rolled-back edges, and bend backwards against the ovary. The petals often are almost completely twisted around to the horizontal. The background color may be creamy white, yellowish, or pale mauve, with reddish brown or reddish purple spots, which may be bright or dull. The lip is magenta-purple with radiating white stripes at the base of the midlobe. The pedicels are quite long in relation to flower size, which produces a very open arrangement of the flowers.

The other important variety, *planilabris,* comes from west Java. The flower color of this variety shows great variability, depending on the locality. See Comber (1982a) for color photos and a description of the two principal varieties.

According to Holttum (1972, 727), there also is a variety or color form called *purpurea,* native to Alor Island (Indonesia), with a background color suffused with purple toward the edges of the sepals and petals, and with deep maroon spotting. This color form seems similar to *Vanda tricolor* 'Vieques,' shown in the American Orchid Society's *Awards Quarterly* (13[4]:255), which received an Award of Merit for the unusual coloring of its flowers.

# 2

# The Crucial Role
# of Chromosomes and Genes

## THE PURPOSE AND PROCESS OF CELLULAR CREATION

The aim of hybridizers of orchids and other living things is to forge new combinations of the inherited genetic material of the parents. They try to produce new combinations of genes that will perpetuate and improve a set of desirable characteristics in the next generation. Some understanding of the transmission of these genetic traits is necessary in order to appreciate the constant references to hybrids and their qualities. Readers interested in more detailed explanations of the mechanisms by which genetic restructuring occurs may consult Raven, Evert, & Eichhorn (1992); Postlethwait and Hopson (1989); and Mehlquist (1974).

Chromosomes and genes provide the means by which parents transmit traits to the cells of their offspring. The nature of the process of cellular creation serves two purposes: (1) it perpetuates the distinguishing features of the species, and (2) it provides a continual diversity of genetic traits in the offspring, which promotes long-term survival of the species. Thus, cellular creation has a twin purpose of providing both stability and change—a very complex and even contradictory pair of goals.

Nature has developed mechanisms for satisfying these two requirements, and they are achieved through two different types of cells: *somatic cells* and *sex cells*. Somatic cells assure programmed development during a plant's life cycle, and sex cells promote diversity in parental offspring.

Somatic cells are also referred to as vegetative or body cells. They form the vegetative tissues of the plant structure, organized into organs such as leaves, stems, roots, and flowers. Each of the somatic cells contains within its chromosomes and genes all the genetic information that is necessary to complete the life cycle, even the potential for reproducing the entire plant. The existence of this capability of totipotency was established in 1958 by Frederick Campion Steward, a botanist

38

and cell biologist. It revolutionized the world of plant cell biology (Sullivan 1993). In the laboratory, it can be demonstrated by tissue-culture cloning techniques; hundreds of plants with identical genetic traits can be produced on order. This process has been used very extensively to produce uniform orchids for the Southeast Asia cut-flower trade, and also on a large scale by producers of plants in the *Cattleya* alliance.

Development of a plant from the embryo in the seed to reproductive maturity is accomplished by division and multiplication of the original somatic cells produced from the fertilized egg. The process of cell division in somatic cells is called *mitosis*. As the newly formed somatic cells complete their development, they differentiate into cells that form the various kinds of tissues and organs of the plant. The process goes on during the entire lifetime of the plant, as new cells and tissues are produced during growth. Cell division by mitosis occurs in tissues called *meristems*. Most often they are found in the apex of new plant growth and in actively growing tips of roots.

Sex cells are produced only in floral tissues. Their development begins during the period of flower-bud formation and is completed only after pollination of the flower takes place. The sex cells develop into *gametes,* which is the name for the mature egg or sperm cells. Gametes are derived from the nuclei of certain somatic cells in the flower by a series of special divisions called *meiosis.* Meiosis has several ordered stages that involve the pairing of chromosomes, reassortment of their genes, and finally production of the actual gametes. The gametes of a plant normally contain half as many chromosomes as its somatic cells, and that halved number is referred to as the *haploid* number.

The functioning of the sex cells, as one might expect from the name, results in procreation. At the same time, the process promotes genetic diversity in the offspring. The egg cells are produced in the ovaries of the flowers; the sperm cells are produced within the pollen tubes that grow after pollination of a flower takes place. After pollination, when the gametes from the pod (female) parent and the pollen (male) parent combine during fertilization, the somatic number of chromosomes is restored, and a new life cycle begins. The genetic information of both parents is combined in the new individual, and that particular information, in the form of the genes on the chromosomes, is then transmitted by mitosis to subsequent cells as they are formed. The fertilized egg is the first cell of the new plant, and its nucleus contains the new somatic set of chromosomes.

## Chromosomes

Chromosomes—the carriers of genetic information—consist of long, thin strands of material called chromatin. These strands are located within the nucleus of each cell and are associated with various proteins and other compounds. The genes are specialized areas of deoxyribonucleic acid (DNA) along the lengths of the strands, somewhat like strings of beads. At the time of cellular division, each single chromosome has two strands of identical chromatin, called chromatids, arranged in the shape of a double helix. The location and spacing of the compounds associated with the double helix produce the genetic code of the cell.

During *mitosis*, the chromosomes, along with their genes, are replicated identically, barring accidents (mutations). The normal, identical replication insures genetic constancy between the old and the new cells of the plant in the various stages of its life cycle. By contrast, in the process of sex cell (gamete) formation during *meiosis*, the genes on the chromatids may be redistributed in an almost infinite number of possible combinations. The randomness of reshuffling and regrouping of the genes during meiosis, followed by the random mating of this egg with that sperm, produces great genetic diversity. That, in turn, promotes adaptability and survival of future generations; "survival of the fittest" can occur only where there is diversity in the offspring of each generation.

## Ploidy and Chromosome Distribution

The term *ploidy* refers to the number of chromosomes in a cell, tissue, or organism. With very few exceptions, all species in the Sarcanthinae subtribe of orchids are *diploids,* meaning that each somatic cell contains two complete sets of chromosomes, one set of which is inherited from each of the parents. The chromosomes of the somatic cells of the diploid plants are grouped into pairs, each member of which is called a *homologue*—one homologue of each pair having come from the male parent and the other from the female parent. All the Sarcanthinae species, with very rare exceptions, have the same number of chromosomes in their cells—38 in the diploid somatic cells and 19 in the corresponding haploid sex cells. So do most of their hybrids. In fact, one reason that these species can be hybridized together is that they all ordinarily have the same respective numbers of chromosomes in their diploid and haploid cells.

The somatic number of chromosomes of a Sarcanthinae diploid plant often is expressed by the formula $2n = 38$, where $n$ stands for the haploid number of chromosomes of a diploid (i.e., 19, the number of chromosomes in each sex cell) and $2n$ indicates the corresponding number of somatic chromosomes (i.e., 38). During a stage of meiosis sometimes referred to as *reduction division,* the somatic number of chromosomes is halved in the process of producing the sex cells. When fertilization occurs with the fusion of egg and sperm, the somatic number of chromosomes is restored in the first cell of the new plant.

The genes attributable to either parent are random and indeterminate. They can vary enormously from one of the offspring to another—within the limits of variability of each species. That is what creates genetic diversity and makes one plant of a species a good breeder for producing fine hybrids, while another plant from the same species population is not.

Although the distribution of the genes cannot be predicted for any given plant, the diploid chromosomes in each somatic cell of the new individual, along with their genes, encompass the entire set of instructions and commands that control every aspect of the plant's development throughout its lifetime. These inherited qualities include both the floral characteristics and the vegetative characteristics, as well as the vigor or physiological characteristics of the new plant. (The culture provided for the plant also affects its growth, and has a great bearing on the actual performance of the plants.)

While the chromosomes and genes are sorted at random during the reduction-division stage of meiosis, one of the partners of each of the homologous pairs of chromosomes may carry more dominant genes than the other partner. It may, as a result, be more powerful in determining certain desirable (or undesirable) traits, and this fact is of great importance to hybridizers. These traits are for the most part invisible when the new plant is inspected, and only the experience of growing it to flowering and of using it in breeding can reveal its full potential. The performance of other members of the grex may be illuminating as a guide, however, and it makes a study of the pedigrees worthwhile. The exceptional and persistent performance of so many plants of *Ascocenda* Yip Sum Wah as stud plants is a case in point.

The randomness of the distribution of the genes also points up why the purchase of seedlings is necessary if growers wish to find new combinations of characteristics in a species or hybrid population. If the hybridizer has used parental plants with known potential for good quality of flowers and plant vigor, the chances are that the same qualities will appear in an appreciable number of the offspring, but there also will be variations from one sibling to the next—some desirable, some undesirable.

## Tetraploids and Other Higher Orders of Ploidy

While vandas and other members of the Sarcanthinae normally are diploid with 38 somatic chromosomes, plants occasionally are found with a greater number of chromosomes. Plants with more than the normal two sets of chromosomes are referred to as polyploid plants. Tetraploids are the most common form of polyploidy in nature, but nonetheless are relatively rare. A tetraploid *Vanda* has cells with double the basic diploid number of somatic and haploid chromosomes—76 of the former and 38 of the latter—often expressed as $4n = 76$, where $n$ is the diploid number (19) of haploid chromosomes. On still rarer occasions, even higher degrees of ploidy are encountered, such as in hexaploids, with six sets of chromosomes ($6n = 114$), or triple the normal diploid number of chromosomes. For two excellent treatments of polyploidy in orchids, refer to Greisbach (1985) and Mehlquist (1974).

Polyploidy can happen through accidents in the processes of mitosis or meiosis. There may be a disruption of the usual divisions brought about by sudden exposure to outside chemicals, by wild fluctuations in temperatures, or even just by chance with other conditions being equal. Or the new cell walls that form to complete the process of division of chromosomes into new somatic or sex cells may not fully develop. In other instances, the chromosomes or the chromatids may not separate after pairing, or may separate unequally. The resulting cells become polyploid, whatever the cause may be.

Deliberate disruptions by hybridizers may also cause polyploidy. Colchicine, an extract from *Colchicum autumnale,* a European and North African crocus, can be used to prevent new cell walls from developing during mitosis, so that the affected somatic cells retain both sets of chromosomes. It is applied while seedlings are still in flask at a very early stage of development.

As ploidy increases, the amount of genetic diversity is enormously enhanced. The somatic cells of the plant become larger, to accommodate the larger number of chromosomes. The flowers of such plants are usually bigger, their substance heavier, and their colors more saturated. The flower sepals and petals are wider than on their diploid counterparts, because the dosage of genes at work in shaping these characteristics is greater. Oddly, the number of flowers on a tetraploid inflorescence generally is slightly less than that of diploid plants, and the flower stalk tends to be shorter and thicker. Despite these characteristics, one cannot with any confidence tell from the relative size of the plant or its flowers whether it is a tetraploid or not. One may suspect that it is, but only an actual chromosome count can make the final determination, and chromosome counting is such a complex laboratory process that it is performed only rarely. Some diploid plants may show such a combination of superior characteristics that one may be led to believe that they are the result of tetraploidy.

There are both benefits and problems in dealing with tetraploid plants. Tetraploids are reputed to be slow growers initially, and consequently take longer to reach maturity. This apparently also is a trait of still higher levels of polyploidy, and is generally considered to be a drawback, a sacrifice for what may be better flowers. By buying the runts in a batch of seedlings, the odds increase of acquiring desirable tetraploid plants in an unknown cross. On the other hand, buying the slow growers also will increase the likelihood of getting genetically defective plants with uneven or partial sets of chromosomes.

Some changes in flower characteristics are not necessarily improvements; the higher gene dosage can work in a perverse direction. For example, no one would want a *Vanda tricolor* flower with more intensely twisted petals. The twisting is pronounced enough in the diploid forms and is not an appealing trait. The challenge to hybridizers is to find those few plants in a population that display improvements in flower form, color, and floriferousness, and no evident worsening in other respects, including vigor of growth. A very good tetraploid can offer the possibility of breakthroughs in hybridizing programs. A favorable assortment of the genes in tetraploids has happened on a number of occasions in the history of breeding hybrid vandas and other orchids and has accounted for many advances.

## Triploids

When a tetraploid is crossed with a diploid plant, the progeny are usually triploids ($3n = 57$), with 38 somatic chromosomes from one parent and 19 from the other. Such plants often combine the best features of both diploids and tetraploids. They are generally vigorous and fast-growing, and they usually bear flowers that are larger than those of diploid plants. The sepals and petals are wider and the colors more vivid. But there may be a small reduction in the number of flowers, and the inflorescence may be shorter. On balance, however, the purchase of some triploids should increase the odds of winning flower awards. Nursery catalogs list such plants either as triploids or with a $4n$ after the name of one parent (to indicate a tetraploid) and nothing after the name of the other parent. (No notation suggests that both parents, and their offspring, are diploids.) If there are any triploids in a group of young seedlings, they are likely to be among the largest plants, not among the runts.

There is one caveat, however, with regard to subsequent hybridizing with triploid plants: they generally do not produce normal sex cells and are therefore sterile. Because of the odd number of chromosomes (57) in the somatic cells of these plants, the homologous chromosomes cannot pair up evenly during the first stage of meiosis, as sex cells begin their formation. This disrupts the orderly process of reduction-division, and the sex cells that are formed will usually be aberrant and not function. They may contain odd pieces of chromosomes or incomplete sets that cannot produce a functional fertilized egg. If fertilization does occur, the offspring will be genetically defective, and the plants will not grow or flower in normal fashion. Such plants are called aneuploids. Triploids are not aneuploids, but any offspring of triploids are likely to be. Occasionally triploids may, just by chance, produce a few functional gametes and they may initiate the formation of a few viable seeds in a pod. Such gametes could have either one or two full sets of chromosomes. There is no assurance that such seed will produce plants that are superior to the triploid parent or, for that matter, to the diploid parent.

## Use of Tetraploids by Thai Breeders

The discovery in Thailand of a few tetraploid ($4n$) plants of *Vanda coerulea* among some nursery seedlings has generated considerable excitement. The potential for using them to produce triploid *Vanda coerulea* plants or hybrids has proven to be excellent. A number of crosses are being made with these tetraploids as pollen parents. The hybridizers have learned that sterility in breeding is not a concern to most buyers of orchid seedlings nowadays, and they hope that customers will be attracted by the possibilities for superior flower shape and color, as well as plant vigor, in the triploid offspring. During an earlier stage, breeders of vandas and ascocendas concentrated their attention on the domestic market for plants, and most Thai buyers wanted plants they could use for subsequent breeding, rather than plants for visual enjoyment. Some Thai breeders still are reluctant to use tetraploids as stud plants, but most of them recognize that the market has changed.

The use of tetraploid plants in Thailand is not limited to those of *Vanda coerulea*. A very small number of 4*n Vanda* hybrid clones are also available, and some of them are being used, especially by Thai breeder Kasem Boonchoo. He has tetraploid forms of *V.* Kasem's Delight, *V.* Gordon Dillon, and *V.* Madame Rattana. Those hybrid vandas have been remarkable parents, even in their diploid forms, and the flowers of their triploid progeny are spectacular. Other breakthroughs resulting from the manipulation of polyploidy by *Vanda* breeders are to be expected.

### Structural Genes and Regulator Genes

Chromosomes have two kinds of genes that govern the physiological and physical traits of a plant: *structural genes* and *regulator genes*. A gene of either kind can be dominant or recessive, or in between.

The structural genes control or influence the basic processes and characteristics of the plant. The regulator genes determine at what stage in the plant's development the structural genes will turn on; and with what degree of intensity the structural genes will operate. The higher the order of ploidy, the larger the total number of structural genes and the greater the likelihood that they will turn on intensely.

The regulator genes may be envisaged as being like an electric dimmer switch combined with an electric timer. When the switch comes on depends on the setting of the timer mechanism, while the intensity of the current depends on the setting of the dimmer switch. For example, a structural gene may control the basic color of the flowers, but the intensity of the color is greatly influenced by how strongly the regulator gene operates on the structural genes it affects. Similarly, another structural gene may control the natural resistance of the plant to certain insects, but how effectively the structural gene actually defends the plant from insect attack depends on when the regulator turns on, as well as on the intensity of its action. If the regulator gene turns on after a majority of the attacking insects have laid their eggs and hatched their larvae, it may be too late to help the plant very much. Timing is all-important.

One aspect of the timing schedule is especially interesting with respect to flower color and color distribution. We know that *Vanda tricolor* and *V. luzonica* each has contributed to the more or less uniformly red and red-speckled *Vanda* hybrids; yet, both *V. tricolor* and *V. luzonica* have discrete bars or blotches of red on their flowers, rather than evenly distributed speckles or minuscule dots of red. What has happened in the hybrids is that the *V. tricolor* and *V. luzonica* structural genes inherited by the hybrids are turned on earlier and more intensely than they are in the species. That results in the red clumps of pigmentation being broken up into much smaller and more uniformly distributed pieces.

Several pigments combine to create the color or colors of flowers. Most of these are affected by the degree of acidity of the cells of the flowers (i.e., their pH). The pH, in turn, is controlled by genes. Some genes can alter the hue and intensity of the pigments; others control the location and distribution of the pigments. With all of the possible combinations of the controlling genes and their effects, one can understand why orchids exhibit such tremendous diversity of hue, intensity, and pattern, and why colors may be formed that are not evident in either of the parents.

One step in the chain of interactions of the genes may affect all subsequent or transitional steps in the expression of the genes. *Alba* forms of flowers may be an example of such an occurrence, where no pigment precursors are formed.

Multitudinous interactions take place among both structural genes and regulator genes during meiosis. In addition, major and minor mutations of genes can create additional effects. Generally, genes are fairly stable and mutate infrequently. This is especially true of major mutations. Occasionally, however, a few genes do mutate spontaneously. The changes produced in a gene by mutation are reproduced in succeeding generations of the gene. As a consequence, individual plants generally are a mosaic of mutated and unmutated genes. Certain genes, referred to as *mutator genes,* not only are prone to mutate themselves but also have a tendency to affect the stability of other genes and to induce them to mutate. Mutations can be triggered by radiation and certain chemical agents. Mutations generally are deleterious, so they rarely are deliberately induced. An exception is the attempt to convert diploids into tetraploids through the use of colchicine.

## SOME PRACTICAL CONCLUSIONS

The siblings of a cross can have an enormous number of possible variations of the genetic material of the parents. The variations in the genotypic traits and qualities may range from invisible or insignificant differences to visible and important effects, and from highly desirable improvements to deplorable faults. Except by extended experimentation and observation of the results, it is almost impossible to anticipate which characteristics of plants chosen as breeding stock will emerge as dominant traits. That is what makes hybridizing as much an art as a science, and why some hybridizers have more successes than others. Chance plays a very large role. By mericloning plants through tissue culture in a laboratory, clones of identical genetic composition can be produced and plant quality maintained; however, the possibility of progress is sacrificed.

The greater the number of different species in the ancestral background of a hybrid offspring, the greater the range and diversity of genetic variations that can be expected to appear from new combinations of chromosomes and genes, including combinations that have never been seen in natural populations. The use of tetraploids as breeding stock in the creation of intergeneric hybrids greatly expands the range of possibilities because of the powerful concentrations of genes in these polyploids. Diversity and differences in the degree of dominance of genetic traits are what create opportunities for hybridizers to develop much-improved plants for the future. The range of possibilities is the source of both their inspiration and their energy.

A remake of a successful previous cross, even with the same clones as parents, has only a small chance of duplicating all the qualities (good or bad) of the offspring of the earlier one. The remake may be better in some respects or worse, but the odds are very small that the progeny will resemble identical twins in all their visible, or phenotypic, characteristics. Look at the differences among human siblings. The same holds true for orchids.

"Blooming out" a very large number of plants of any given cross is highly desirable—even essential—in an extended program of selective breeding. A small sample of plants may easily miss the rare exceptional specimen possessing the phenotypic qualities the hybridizer is seeking to find and incorporate in the following generation of plants.

The potential of a plant as a parent cannot be determined from one effort or with one mate. Several efforts, including using more than one mate, are needed. A further complication is that it makes a difference whether a plant serves as the pod (female) parent or the pollen (male) parent.

The traits most favored by orchid enthusiasts—traits such as better arrangement of the blossoms on the inflorescence—are not necessarily the ones Nature is seeking to achieve. Nature is more concerned with attractiveness to pollinators, resistance to insects and disease, and other qualities relevant to survival of the species. Plants with those qualities will survive and multiply in the wild. Plants with prettier flowers may not, unless they happen to have strong survival traits as well. If artificially propagated species are to be introduced into the wild to replenish depleted native populations (which up to now has not been done to any significant extent, but may become more frequent as conservation programs expand), it is important that the plants not be the result of several generations of selective breeding for show specimens. Instead, they should contain as much of the original genetic diversity of dominant and recessive genes as possible, and any selective breeding should be directed toward strengthening traits beneficial for survival, such as vigor of growth.

While the process of meiosis creates tremendous genetic diversity in order to ensure adaptability and survival, "superior" genes do tend to beget "superior" genes in their progeny; superior qualities thereby are transmitted by parents to at least some of their offspring. Moreover, if the superior qualities are present in both of the parents, there is a fair chance that they will appear in still further improved form in the offspring. Since good genes are so important, seek out plants that give some expectation of having inherited them. Always remember, as John Wolcot remarked some 200 years ago: "You cannot make, my Lord, I fear, a velvet purse of a sow's ear."

# 3

# The Species Ancestry
# of *Vanda* Hybrids

## THE DETERMINANTS OF HYBRID ANCESTRY

The ancestry of today's superb *Vanda* hybrids reflects the standards and goals of the hybridizers who developed them, and the preferences of the public they were seeking to please. In the early stages of large-scale hybridization of vandas, which occurred in Hawaii during the mid- and late 1940s and through the 1950s, breeders generally adhered to the broad English standards of flower judging. These gave strong preference to large flowers with round, flat shape. Taking this orientation as their guide, the Hawaiians then developed specific parameters to be applied to *Vanda* hybrids, and to other orchids grown in Hawaii as well. The Thai breeders, who dominated hybridizing in the 1970s and 1980s, retained the Hawaiian standards unchanged.

In the late 1980s and early 1990s, some limited interest developed in the United States and abroad in exploring or reexploring the potential of primary crosses of species within the *Vanda* genus and of crosses of *Vanda* and *Ascocenda* (*Vanda* × *Ascocentrum*) hybrids with species of other vandaceous genera. Some of the resulting combinations do not have large, round, flat flowers but do have other desirable horticultural attributes, such as small, compact plants that can be grown under lights or in small greenhouses, interesting flower shapes, new color combinations, and fragrance—attributes that, for the most part, carry little or no weight in the judging regimes of the American Orchid Society or the Royal Horticultural Society of Great Britain. The criteria and values applied by these two societies are largely accepted and followed throughout the world.

It is premature to speculate to what extent recent innovative efforts, especially when they depart from the established norms of "beauty," will succeed in gaining broad acceptance and popularity. As of the early 1990s, however, there has not been any revamping of the traditional standards applied in evaluating flower quality under the judging regimes of the American Orchid Society, the Royal Horticultural Society of Great Britain, and elsewhere.

The standards described in this book are the conventional and still widely accepted ones. Nevertheless, innovative hybidizing should be encouraged, regardless of whether the probable outcome will be larger or rounder or flatter flowers. (The subject of flower quality is treated in detail in Chapter 4.) The development of today's much-applauded hybrids has been shaped primarily by the historical perception of the "ideal"; in fact, the perceived degree of departure from the ideal, in the eyes of the judges, is the explicitly stated cornerstone of the formal judging system of the American Orchid Society. It also is the foundation of the point-scoring system of the German Orchid Society.

Among the *Vanda* species, *Vanda sanderiana*, which was introduced into cultivation in England in 1882, has represented the historical and the current ideal. It is not surprising, therefore, that it is dominant in the ancestry of what are considered to be the best *Vanda* hybrids of the past and present. In the 1990s, however, this dominance of *V. sanderiana* came to be shared with *V. coerulea*, the other large-flowered *Vanda* species, especially as breeders developed rounder and flatter forms of that species. Although there are about 50 *Vanda* species, today's popular hybrids are descended primarily from these two.

Of the two, *Vanda sanderiana* appears on the family tree more often than *V. coerulea*, and it has had a greater degree of influence on the standards used in judging *Vanda* hybrids. Selective breeding of *V. sanderiana* has produced some remarkably good cultivars, with flowers that differ little in appearance from those of some of the best similarly colored hybrids. The hybrids, however, have other good qualities that *V. sanderiana* lacks. Thus, hybridizers have progressed in two directions simultaneously with *V. sanderiana*: first, they have produced some vastly superior cultivars of that species; and, second, they have produced large, flat, full-shaped hybrids that bloom at an earlier age, more frequently, and in a wider range of colors.

Most *Vanda* (and *Ascocenda*) hybrids have more *V. sanderiana* ancestors than *V. coerulea* ancestors, but the latter species appears much more frequently on the family tree of most hybrids than does any other of the *Vanda* species, primarily because the flowers of the other species are considerably farther removed from the established standard of beauty.

The use of other species has been limited, for the most part, to compensating for some of the limitations of flower color inherent in the *Vanda sanderiana* and *V. coerulea* gene pools. The other species in the background of the most highly regarded hybrids have not helped to satisfy conventional notions about desirable shape, size, arrangement, or number of flowers. Introducing additional *Vanda* species holds little or no promise of help in this regard; in fact, that route was tried and abandoned long ago. If the traditional standards of judging are relaxed and broadened as time goes by, however, the experimentation with other species of vandas may be resumed on a larger scale. Fashions in orchids, like other fashions, are not immutable.

## TRAITS OF *VANDA SANDERIANA* AND *VANDA COERULEA*

*Vanda sanderiana*, the principal species incorporated in *Vanda* hybrids, serves as the prototype for the conventional standards applied to *Vanda* hybrids. In its superior forms (see Plate 23), *V. sanderiana* has flowers that are fairly round, flat, and attractively colored. Flower size is large, generally measuring about 10 cm (4 in) across. Usually the flowering period is late summer or early autumn, although in cultivation a plant may come into bloom at other than the normal time, as so often happens with *Vanda* species transplanted from their native habitat.

The inflorescences of *Vanda sanderiana* typically carry 8–12 flowers. Exceptional plants with more flowers have been recorded, and awarded cultivars tend to have 12–15 flowers. The individual flowers have good substance and last for several weeks. Mature plants often have two or more inflorescences when in bloom. They also have a tendency to send out vegetative shoots from the base of the plant. When these shoots bloom at the same time as the mother plant, the floral display is striking.

On well-grown plants, the flowers hold themselves above the foliage. They encircle the flower stalk evenly, forming a complete cylinder of flowers—an eminently desirable quality not shared by the flowers of some other *Vanda* species, which are characterized by a rather higgledy-piggledly arrangement. Moreover, the stout pedicels (the stems attaching the individual flowers to the flowering stalk) are well sized for displaying the flowers, unlike the pedicels of many other *Vanda* species, which often are disproportionately long and thin and do not permit the individual flowers to be displayed to best advantage.

Another attraction of *Vanda sanderiana* is that it is found with several gradations of flower coloration, although, with the exception of the *alba* form, the differences really are very minor variations on a theme rather than distinctly different color forms. For outstanding photographs of the color variations, see Valmayor (1984, 2:242–251).

All the color forms have a marked contrast between the upper half of the flowers, consisting of the petals and the dorsal sepal, and the lower half, consisting of the two lateral sepals. The upper half is much paler than the lower half. Except in the *alba* form, the petals and the dorsal sepal range in color from an almost creamy white or pale pinkish white to a rosy lilac, and they have irregular, scattered speckling of brownish red or purple near the base. The lateral sepals are much more darkly colored, usually a tawny yellow or chartreuse background color overlaid with heavy veining or tessellation. The veining may range from a cinnamon to a chocolate-brown color. On some plants, the veining has been described as being purplish, or purplish brown. On most plants, the veining is fairly crisp and well defined, but on some cultivars the color suffuses into the adjacent spaces.

In the *alba* form of *Vanda sanderiana* (Plate 26), the petals and the dorsal sepal are nearly white, and the basal spotting is the same shade of green as a Granny Smith apple. The lateral sepals also are apple-green, and the tessellation is quite subdued. The flowers of the *alba* form tend to be somewhat smaller than those of the standard color form, and they also are less numerous. It should be noted that these traits are typical of *alba* forms of orchids; however, these inferiorities are lost when *alba* forms are used to produce hybrids—the latter have full vigor. The *alba* form of *V. sanderiana* has been used in making hybrids with yellow and greenish yellow flowers. In making such hybrids, it is desirable to use a *V. sanderiana* with flowers lacking intense tessellation or deeply colored lateral sepals, and that mandates using the *alba* form.

With all its desirable qualities, it is easy to understand why *Vanda sanderiana* set the standard for *Vanda* hybrids and why, until very recently, it reigned supreme among the *Vanda* species after its introduction into England in 1882. Hybridizers soon came to realize that *V. sanderiana* has an exceedingly rich pool of genes they could put to good use. Their efforts were facilitated by the soon-discovered fact that *V. sanderiana* breeds easily and grows quite vigorously. The early English and European growers found *V. sanderiana* much easier to grow than *V. coerulea*, probably because it took them a while to realize that *V. coerulea* likes cool nights, whereas *V. sanderiana* prefers constant warmth, which is what the pioneer English and European growers thought *all* vandas required.

Notwithstanding its many considerable virtues, *Vanda sanderiana* has some shortcomings, most of which can be overcome by crossing it with *V. coerulea*. Its petals, with rare exception, are rather small in relation to the size of the lateral sepals, and therefore to the overall size of the flower. The petals tend to slant upward, which further exacerbates the first fault by revealing more of the lateral sepals. Moreover, the petals are rather long in relation to their width; greater roundness would give a more pleasing impression of balance. The relative narrowness of the petals, even combined with their upward-tilting trait, normally prevents them from coming as close to each other where they overlap the dorsal sepal as one would like. Extensive line-breeding of highly selected forms of *V. sanderiana* has produced an occasional exceptional specimen with better-shaped and larger petals, but the flower shape generally falls somewhat short of that of the best hybrids that, in other respects, resemble *V. sanderiana*.

Another problem with *Vanda sanderiana* is that it blooms only once a year, and, to make matters worse, it usually does not begin blooming until the plant is six or seven years old. Again, some of the best line-bred plants are an improvement in this latter respect and bloom at a somewhat earlier age.

A further undesirable characteristic is that the flowers tend to be borne in a bunch at the top of the flowering stalk—a sort of popsicle look—instead of being attractively spaced along a longer inflorescence as they are on *Vanda coerulea*. The line-bred plants are not notably improved in this respect.

In addition, the colors of *Vanda sanderiana* frequently are a bit dull and lack intensity or saturation. Unfortunately, this trait tends to be passed on to its immediate hybrid progeny. Also, despite the existence of minor variations in color, the range is narrow and excludes much of the spectrum.

If ever a marriage of two species was made in heaven, it was the mating of *Vanda sanderiana* with *V. coerulea*, which produced *V.* Rothschildiana (see Plate 28). The cross was made by M. Chassaing, the chief gardener of the Rothschild chateau Ferrieres-en-Brie, in Seine & Marne, France, and registered in 1931. The strengths of *Vanda coerulea* almost completely counterbalance the weaknesses of *V. sanderiana* and vice versa. A superior *V. coerulea* plant blooms two to four times a year, and begins to bloom at a very early age. The flowers of such plants are well spaced on a long inflorescence bearing at least as many flowers as one normally finds on a clone of *V. sanderiana*, and quite often more. On fine cultivars, the flowers tend to be as large as those of *V. sanderiana*. On superior clones of *V. coerulea*, the dominant color ranges from a pure sky-blue to a deep violet. Actually, the background color is white, or white tinged with a purplish cast, but the intense, uniform, blue or violet tessellation produces an overall impression of blue or violet. The petals on good specimens are round and are well proportioned in relation to the size of the flower.

By combining *Vanda coerulea* with *V. sanderiana*, it was possible to introduce blue-purple pigmentation and tesselation into the flowers of *Vanda* hybrids, while still preserving the best qualities of both parents. *Vanda* Rothschildiana is the prime example.

Not every cultivar of *Vanda coerulea* has all the virtues described, or to full degree, but a number of them do. Even some of the collected specimens taken to England in the late 19th century were exceptionally good, judging from descriptions in the *Orchid Review,* the leading English orchid journal, in publication since 1893. In the mid-1960s, when the people of Thailand began to take orchids seriously, it did not take growers there long to acquire superior specimens from jungle plants collected in northern Thailand and Burma. They used these to produce even better ones. By about 1975, some excellent plants of *V. coerulea* were available to hybridizers in Thailand and were being used in breeding programs utilizing both *V. coerulea* and improved forms of *V. sanderiana*.

Plate 1. *Vanda coerulea* 'Orchidgrove Lois', AM/AOS.

Plate 2. *Vanda dearei.*

Plate 3. *Vanda denisoniana.*

Plate 4. *Vanda lamellata.*

Plate 5. *Vanda luzonica.*

Plate 6. *Vanda merrillii.*

Plate 7. *Vanda roeblingiana.*

Plate 8. *Vanda tessellata*.

Plate 9. *Vanda tricolor*.

Plate 10. *Ascocentrum ampullaceum.*

Plate 11. *Ascocentrum curvifolium*.

Plate 12. *Ascocentrum miniatum.*

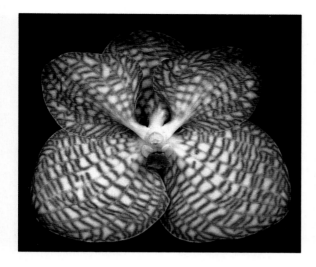

Plate 13. *Vanda coerulea* 'Lois Grove', FCC/AOS. This plant culminates several generations of highly selective breeding to improve the form and color of the species. The results are far superior to the jungle-collected plants. Photograph by Charles Marden Fitch.

Plate 14. *Vanda coerulea* 'Grove's Delight', an extraordinarily dark color-form of the species; the shape, number and arrangement of the flowers is not outstanding.

Plate 15. *Vanda coerulea* 'Orchidgrove', CHM/AOS, was awarded for its rare, attractive pink color. It has been little used in hybridizing—the color genes probably are recessive. Photograph by Charles Marden Fitch.

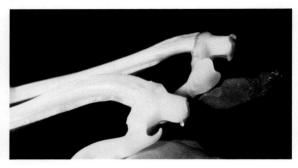

Plate 16. *Vanda coerulea* f. *alba.* This exceedingly rare pure white form should not be confused with the "washed-out" whitish specimens with a purple lip; those are not highly regarded. Photograph by Charles Marden Fitch.

Plate 17. Below: side lobe of the lip of *Vanda coerulea,* showing the characteristic recurved, sharply hooked feature of the tip. Above: a hybrid with ¾ *V. coerulea* ancestry, showing the shorter, blunter tip produced when other species are introduced. Photograph by William Smiles.

Plate 18. This plant received an FCC/AOS as a *Vanda coerulea,* but the shorter, less-hooked tips of the side lobes, and the shape of the petals, are very different from those of a genuine *V. coerulea;* it is an attractive hybrid. Photograph by Ernest Walters.

Plate 19. *Vanda* Sansai Blue (*V.* Crimson Glory × *V. coerulea*). another striking *V. coerulea* hybrid, in this instance made with a tetraploid *V. coerulea* parent. The hybrid displays ideal *V. coerulea* shape. Photograph by Charles Marden Fitch.

Plate 20. *Vanda cristata* 'Kerry Slora', AM/AOS. The distinctive lip is the dominant feature of this small-flowered species, and it tends to carry through to its primary hybrids. Photograph by Joseph E. Buggie.

Plate 21. *Vanda lamellata* var. *boxallii* 'J.E.M.', AM/AOS. Note the vivid coloring of the inside part of the lateral sepals. For many years, this variety was considered to be an independent species, and was much favored by the early hybridizers of vandas. Photograph by Bob Smith.

Plate 22. *Vanda parviflora* (= *V. testacea*) 'Lauray', HCC/AOS. Note the twisted petals and long pedicels characteristic of many *Vanda* species. Photograph by Charles Marden Fitch.

Plate 23. *Vanda sanderiana* 'Orchidgrove', a superb artificially propagated plant from R. F. Orchids. Its parents are *V. sanderiana* 'Coral Reef', AM/AOS, and another highly selected *V. sanderiana*. Two of the grex have received FCC awards from the AOS. Photograph by Charles Marden Fitch.

Plate 24. *Vanda* Joan Viggiani 'Orchidgrove', AM/AOS. A hybrid whose flowers closely resemble those of *V. sanderiana*, but with different lip structure (see Plate 25). Photograph by Charles Marden Fitch.

Plate 25. *Vanda* Joan Viggiani, top, has a spur on the back of its lip, a characteristic of all vandas other than *V. sanderiana*, below. Largely because of absence of a spur, some taxonomists place *V. sanderiana* in a monotypic genus, *Euanthe*. Photograph by William Smiles.

Plate 26. *Vanda sanderiana* f. *alba* 'Eastwind', HCC/AOS. The *alba* color form has been used extensively in making *Vanda* and *Ascocenda* hybrids with yellow or chartreuse flowers. Photograph by Charles Marden Fitch.

Plate 27. *Vanda spathulata* (= *Taprobanea spathulata*) 'Genevieve', AM/AOS. Whenever this species is combined with any other species or hybrid in the Sarcanthinae subtribe, its influence on color is dominant. It is reported to be a hexaploid. Photograph by Philip Sidran.

Plate 28. *Vanda* Rothschildiana 'Robert', AM/AOS *(V. coerulea × V. sanderiana)*. The flowers of this cultivar set a new standard for this, the most highly awarded *Vanda* hybrid grex. It has the best qualities of both the parent species. Photograph by Bob Smith, courtesy of R. F. Orchids.

Plate 29. *Vanda* Fuchs Oro (*V.* Pong Tong × *V.* Bangkhunsri) has clear golden color with minute cinnamon dots and some tessellation, both inherited from *V. sanderiana* ancestry and hard to avoid in yellow hybrids. Photograph courtesy of R. F. Orchids.

Plate 30. An unidentified *Vanda* hybrid with chartreuse flowers. The few flowers and rather rough surface texture are typical of many efforts to produce yellow-flowered *Vanda* hybrids. Most have an overlay of subdued rust-colored streaks and dots.

Plate 31. *Vanda* Rasri Gold 'Orchidgrove', AM/AOS (*V.* Kultana Gold × *V.* Thananchaisand). Note the uniform yellow color of all parts of the flower including the lip, the round, flat form, and smooth texture, which combine to make this perhaps the best yellow *Vanda* to date. Photograph by Charles Marden Fitch.

Plate 32. *Vanda* Fuchs Delight (*V.* Gordon Dillon × *V.* Kasem's Delight) is an outstanding hybrid, registered by R. F. Orchids and made at least several times by others, too. It comes in the same color forms as its parents. Note the large, wide petals and the full, round shape. Photograph by Charles Marden Fitch

Plate 33. *Vanda* Kasem's Delight 'Krachai', HCC/AOS (*V.* Sun Tan × *V.* Thosphol), a Kasem Boonchoo hybrid. Like *V.* Gordon Dillon, it comes in red, pink, and purple color forms, with varying degrees of speckling and venation. No other modern *Vanda* hybrid has produced as many awarded progeny. Photograph by Charles Marden Fitch.

Plate 34. *Vanda* Gordon Dillon (*V.* Madame Rattana × *V.* Bangkok Blue), an unidentified cultivar. This cross, made by Kasem Boonchoo of Bangkok, comes in several color forms. Photograph by Charles Marden Fitch.

Plate 35. *Vanda* Gordon Dillon, another unidentified cultivar and another color form of this famous hybrid, which has been used as a parent by many hybridizers. Photograph by Charles Marden Fitch.

Plate 36. *Vanda* Antonio Real (*V.* Fuchs Delight × *V.* Danny German), an ideally shaped *Vanda,* by current standards. Note the perfect proportions of the sepals and petals, and their attractive coloring. Photograph courtesy of R. F. Orchids.

Plate 37. *Vanda* Thong Chai (*V.* Ponpimol × *V.* Kasem's Delight), a beautifully tessellated red progeny of *V.* Kasem's Delight. Photograph by Charles Marden Fitch.

Plate 38. *Vanda* Jason Robert Fuchs (*V.* Thospol × *V.* Adrienne). The markings and vibrant red color of this flower, along with its wide petals and round, flat shape, make it exceptional. V. Thosphol also was a parent of *V.* Kasem's Delight. Photograph by Charles Marden Fitch.

Plate 39. *Vanda* Deva 'Orchidgrove', AM/AOS (*V.* Crimson Glory × *V.* Thospol), made by Charungraks Devahastin. The uniform deep-violet color with little visible venation is fairly rare. Photograph by Charles Marden Fitch.

Plate 40. *Vanda* Grove's Pleasure 'Orchidgrove'. Crossing *V.* Deva 'Orchidgrove', AM/AOS (Plate 39), with *V. corulea* 'Lois Grove', FCC/AOS (Plate 13) produced vivid, full, flat, round, tessellated flowers on plants that bloom at a very early age. Photograph by William Smiles.

Plate 41. *Vanda* Michael Coronado 'Bojote', AM/AOS, a cross of *V.* Deva (Plate 39) and *V.* Yen Jitt. The color is inherited from the pollen parent. The shape is excellent. Photograph by Bob Smith, courtesy of R. F. Orchids.

Plate 42. *Vanda* Keeree (*V.* Dawn Nishimura × *V.* Patou), registered in 1980, has huge, flat, bicolored flowers, reminiscent of *V. sanderiana* but more vibrantly colored. It has parented some exceptional hybrids, such as *V.* Faye Bennett, *V.* Keeree's Delight, and *V.* Fuchs Cheers.

Plate 43. *Vanda* Fuchs Violetta (*V.* Kretcant × *V.* Kasem's Delight), a bright pink flower with round, flat shape—a much desired combination. Photograph courtesy of R. F. Orchids.

Plate 44. *Vanda* Carmen Coll (*V.* Varavuth × *V.* Faye Bennett). The distinctive dark pink, crisp tessellation greatly enhances the blush-pink background. Photograph by Bob Smith, courtesy of R. F. Orchids.

Plate 45. *Vanda* Patricia Lee (*V.* Ellen Noa × *V.* Gertrude Miyamoto). This unusually colored flower suggests the influence of the distant *V. tricolor* ancestry of *V.* Gertrude Miyamoto. Its only other ancestral species are *V. dearei* and *V. sanderiana*. Photograph by Charles Marden Fitch.

Plate 46. The dorsal sepals and the petals of this unidentified *Vanda* hybrid are pinched and re-curved at their base—a common fault of first-generation *V. coerulea* hybrids. The inflorescence has too few blossoms and a bunched arrangement.

Plate 47. An unidentified *Vanda* hybrid with excessively long pedicels, which cause the flowers to tilt downward and make them harder to appreciate. The flowers also are spaced too far apart.

Plate 48. A *Vanda coerulea* whose flowers are spaced too far apart. The twisting of the petals is typical of jungle-collected *V. coerulea* and of most man-made ones as well. Untwisted petals, prized by orchid judges, really are unnatural in this species.

In terms of the established standards, the one significant "fault" of *Vanda coerulea* is its twisted petals. This, however, has not been an insurmountable problem with its hybrid progeny; first, because the remarkable flatness of the petals of *V. sanderiana* seems to be a rather dominant trait when that species is crossed with *V. coerulea*; and, second, because intensive line-breeding of superior forms of *V. coerulea* has diminished this trait somewhat.

Mating *Vanda coerulea* with *V. sanderiana* to produce *V.* Rothschildiana, and then manipulating the resultant gene pools again and again, produced some excellent hybrids from the very beginning. Among these were *V.* Onomea (*V.* Rothschildiana × *V. sanderiana* ) in 1948, *V.* Rose Davis (*V.* Rothschildiana × *V. coerulea*) in 1951, and *V.* Jennie Hashimoto (*V.* Onomea × *V. sanderiana*) in 1954. Indeed, *V.* Rothschildiana appears somewhere in the background of most of today's *Vanda* hybrids.

## THE CONTRIBUTIONS OF OTHER SPECIES

What, then, did hybridizers hope to achieve by introducing other *Vanda* species whose flowers had neither large size, nor good shape, nor flatness of form? True, some of them have the desirable qualities of frequent blooming and early blooming, but *V. coerulea* provides these in good measure, too. The possibility of transmitting fragrance—a pleasant quality of a number of *Vanda* species—may have been an attraction, although, in many orchid genera, fragrance seems to be a recessive trait, so it is doubtful that hopes on this count were high.

The primary objective of early hybridizers was to create a greater range of vibrant colors and interesting color combinations. *Vanda sanderiana* tends to impart dullness or muddiness of color to its hybrids. *Vanda coerulea*, at least in the first generation, has great difficulty in overcoming this fault, as is evident in many cultivars of *V.* Rothschildiana (*V. coerulea* × *V. sanderiana*). *Vanda* Rothschildiana's flowers usually are a dull grayish blue, despite the brightness of the blue or violet pigmentation present in the tessellation of superior color-forms of *V. coerulea*. This tendency toward dullness is not invariable, but it occurs with sufficient frequency to be a significant concern.

The problem, however, is that the flowers of most of the other *Vanda* species are yellowish or greenish brown, in addition to having poor shape and arrangement and smaller size. While *V. luzonica* (Plate 5), *V. merrillii* (Plate 6), and *V. tricolor* (Plate 9), along with a very few other exceptions, are strikingly colored, most *Vanda* species are hardly a hybridizer's delight in their own right, as may be deduced from the descriptions in Chapter 1.

Nevertheless, hybridizers, whether from experience with other genera or from intuition, put their faith in one of the fundamental laws of hybridization. That law, as succinctly stated by Henry M. Wallbrunn (1984, 380), is that "gene combinations give traits outside the range of the two parent species." So, being eternal optimists, hybridizers could hope for some spectacular results. Much of their optimism was well founded. By adding some other species to *Vanda sanderiana–V. coerulea* combinations, they soon produced pleasing new colors and color patterns not previously found in any of the *Vanda* species. (How this is possible is explained in Chapter 2.)

Magnificent *Vanda* hybrids with saturated red and pink colors now exist—colors that cannot be attributed to *Vanda sanderiana* and *V. coerulea* ancestry alone. Plants with flowers of more uniform colors, and with little tessellation, have been developed, and ones with more speckling and other configurations of color. Glistening, crystalline texture is evident as well.

It is foolish to try to attribute each of the new or improved qualities to particular other species in a hybrid ancestry, because, more often than not, what we observe is not a direct transmittal of pigmentation from a parent but rather the result of a complex interaction within the hybrid's entire genetic inheritance. This point cannot be overemphasized. Hybridization affects when and with what intensity the regulator genes "turn on," and that is not predictable, but experience sometimes can provide a guide. For example, empirical evidence supports the conclusion that the genes of both *Vanda tricolor* and *V. luzonica* generate interactions with the color-controlling genes of *V. sanderiana* and *V. coerulea* in ways that often produce flowers with suffused, saturated, and sparkling red or pink pigmentation, as well as flowers with striking speckling, not only in red hybrids but in blue and purple ones as well. Yet these combinations are not found in any of the foregoing four species.

By adding *Vanda tricolor* (Plate 9) or *V. luzonica* (Plate 5) genes, the range of *Vanda* flower colors from pink through purple, and their patterns, were greatly broadened, and clarity of color was improved. Oddly enough, there is no apparent difference in the range of colors and color patterns between groups of crosses both of which have similar proportions of *V. sanderiana* and *V. coerulea* ancestry, but with one group also having *V. tricolor* in its background while the other has *V. luzonica*.

Another conclusion derived from experience is that *Vanda dearei* plants produce yellow-flowered hybrids if there is a *V. dearei* ancestor as close as a great-grandparent. The color-controlling genes of *V. dearei* plants evidently are quite dominant. Unfortunately, *V. dearei* (Plate 2) also transmits its less appealing characteristics: paucity of flowers; indifferent shape; short inflorescences; large plant size; and intolerance of cold. Yellow-flowered *Vanda* hybrids, being heavily influenced by their *V. dearei* ancestry, generally exhibit the same unwelcome traits as that species. Vandas with red, pink, blue, or bicolored flowers have much less *V. dearei* ancestry than do yellow-flowering hybrids.

When hybridizers attempt to preserve yellow flower color but to eliminate the faults of Vanda *dearei* by increasing the proportionate share of *V. sanderiana* genes, the result almost always is a reintroduction of both tessellation and traces of flower colors other than yellow (see Plate 29). Moreover, in the process, the surface of the petals and sepals often displays a somewhat rough texture (see Plate 30). Some fine pure-yellow vandas do indeed exist (see Plate 31), but the chances of obtaining one from any group of yellow-bred seedlings is small. Undoubtedly, the situation will improve in the future. For the present, however, those who want long inflorescences with many large, round, flat, well-arranged flowers are advised to seek vandas bred to produce colors other than pure yellow or chartreuse.

A third example where lessons from experience are widely known is *Vanda tessellata* (Plate 8). It would seem to have interesting possibilities for transmitting color, given the variations in the colors of the flowers of the species, and its attractively colored lip. Its open shape and its narrow, wavy petals, however, are deterrents to much use of this species in the production of standard-type *Vanda* hybrids, because the undesirable elements of flower shape are difficult to "breed out." A few hybridizers, however, are continuing to experiment with *V. tessellata*. One interesting novelty-type hybrid from such efforts is *V.* Golden Doubloon (*V. denisoniana* × *V. tessellata*). A "novelty-type" is a plant with a flower shape quite different from the large, round, flat standard expected of plants with predominantly *V. sanderiana* and *V. coerulea* ancestry.

## THE INTERNATIONAL RECORD OF *VANDA* HYBRIDS

Information on ancestry is essential to any attempt to understand and benefit from the evolution of today's advanced hybrids. Orchid growers are very fortunate in having a published international record, extending back nearly a century, of all known orchid hybrids, their parents, date of the grex, the hybridizer, and, generally, the hybridizer's address. We owe this good fortune to Fred. K. Sander of the orchid firm of Sander & Sons, which was located in St. Albans, England.

In 1895, Sander began to record, in what his nephew called a series of "inexpensive notebooks," the names and other pertinent available information about all existing orchid hybrids, and new ones as they were made. This was knowledge sorely needed by the orchid community, especially in order to prevent the chaos that would exist if the same hybrid made by different persons were to appear on the market, and hence in collections, under different grex names. That sometimes had been the case.

The pleas to share his information induced Sander to print his list in 1905. Subsequent addenda were issued, some years apart. Later, his nephew, David F. Sander, carried on the work, which developed into a more formal system of registration.

In 1960, the Royal Horticultural Society, London, agreed to assume the by then arduous duties of International Registration Authority for Orchid Hybrids, and to follow the provisions of the International Code of Nomenclature for Cultivated Plants. The International Registration Authority for Orchid Hybrids was established by the Commission for Nomenclature and Registration, which has appointed various international registration authorities for hybrids of various plant genera. That commission, in turn, is one of a number established by the International Society of Horticulture, located in the Hague, Netherlands, to deal with various aspects of horticulture. The Royal Horticultural Society began to carry out these duties from January 1, 1961, onward, including periodic publication of new registrations. In the interest of continuity, and in recognition of the Sander family's great contribution in the past, the title *Sander's List of Orchid Hybrids* was retained. The information on the parentage of *Vanda* and *Ascocenda* hybrids in this book comes from that source.

By year-end 1994, *Vanda* hybrids had been made with 28 different strap-leaf *Vanda* species, of those listed as such in *Sander's List* (compare with Christenson's list of the genus *Vanda*, in Appendix B), and *Vanda* species had been used as a direct parent to make more than 500 different *Vanda* hybrids. Hybrids with any ancestry of terete species are not included in this count.

## Relative Use of Individual *Vanda* Species

*Vanda* hybridizers began their work by making primary hybrids, using two species of vandas as parents. It took a while before there were enough hybrids available for breeding programs to be developed very far. The first two recorded strap-leaf *Vanda* hybrids were registered in 1919, apparently. One was *V.* Gilbert Triboulet, made in France by Jean Gratiot. Its parents were *V. coerulea* and *V. tricolor*. The other was *V.* Tatzeri, made at the Prague Botanical Gardens. Its parents were *V. sanderiana* and *V. tricolor*.

The next two hybrids were registered in 1921. One was *Vanda* Herziana (*V. coerulea* × *V. suavis*); however, since *V. suavis* now is considered to be simply a variety of *V. tricolor*, *V.* Herziana is the same cross as *V.* Gilbert Triboulet as far as the International Registration Authority for Orchid Hybrids is concerned. *Vanda* Herziana was made by "P. Herz," whose location is not given in *Sander's List*. Henderson and Addison (1956, 134), however, state that Herz raised the hybrid in Hawaii. The other *Vanda* hybrid registered in 1921 is *V.* Mariannae (*V. denisoniana* × *V. tricolor*), made at the Prague Botanical Gardens.

Tables 3-1 and 3-2 present a tabulation of the *Vanda* hybrids recorded up to January 1, 1946. Table 3-1 includes only primary hybrids. Table 3-2 records other hybrids; in other words, the next step in the evolution of the advanced hybrids of today. The two tables are of historical interest because they reveal where the pioneering work on *Vanda* hybrids began and, equally important, which species attracted the early hybridizers.

*Vanda* hybridizing was very slow in getting under way, compared with hybridizing within such other orchid genera as *Cattleya, Cymbidium, Odontoglossum,* and *Paphiopedilum.* Hybridizing within those genera started earlier and developed more rapidly (see *Sander's List of Orchid Hybrids* for the period prior to 1946).

It is astonishing that the English had little evident interest in hybridizing vandas. English orchidists were in the forefront in producing hybrids in other orchid genera. The lack of interest in vandas cannot be attributed entirely to climate. The climate in France, Czechoslovakia, and Germany, where the earliest *Vanda* hybrids

Table 3-1. Strap-leaf *Vanda* primary hybrids registered up to January 1, 1946
(from *Sander's List of Orchid Hybrids*).

| YEAR | HYBRID NAME | SPECIES PARENTAGE | HYBRIDIZER | PLACE |
|---|---|---|---|---|
| 1919 | Gilbert Triboulet[a] | *coerulea × tricolor* | Jean Gratiot | France |
| 1919 | Tatzeri[a] | *sanderiana × tricolor* | Prague Bot. Gdns. | Czechoslovakia |
| 1921 | Herziana[a] | *coerulea × tricolor (suavis)* | P. Herz | N.A. |
| 1921 | Mariannae | *denisoniana × tricolor* | Prague Bot. Gdns. | Czechoslovakia |
| 1927 | Kupperi | *lamellata (boxallii) × sanderiana* | Munich Bot. Gdns. | Germany |
| 1928 | Burgeffii[a] | *sanderiana × tricolor (suavis)* | Munich Bot. Gdns. | Germany |
| 1928 | Boschii | *luzonica × tricolor* | Munich Bot. Gdns. | Germany |
| 1931 | Rothschildiana | *coerulea × sanderiana* | Chassaing | France |
| 1934 | Mem. T. Iwasaki | *dearei × tricolor* | Prince Shimadzu | Japan |
| 1943 | Manila | *luzonica × sanderiana* | Rapella Orch. Co. | California |
| 1944 | Helen Adams | *dearei × tricolor (suavis)* | Ernest de Saram | Ceylon |
| 1944 | Paki | *cristata × tricolor* | H. C. Shipman | Hawaii |
| 1945 | Flammerolle | *coerulea × luzonica* | Vach. & Lecoufle | France |
| 1945 | Jill Walker | *coerulea × lamellata (boxallii)* | F. C. Atherton | Hawaii |
| 1945 | Kahili Beauty | *lamellata × tessellata (roxburghii)* | B. Tanaka | Hawaii |
| 1945 | Kapoko | *lamellata (boxallii) × tricolor* | H. C. Shipman | Hawaii |
| 1945 | Lester McCoy | *coerulea × dearei* | J. R. Cummins | Hawaii |
| 1945 | Loke | *lamellata (boxallii) × luzonica* | H. C. Shipman | Hawaii |
| 1945 | Caroline J. Robinson | *tessellata (roxburghii) × tricolor* | H. C. Shipman | Hawaii |
| 1945 | Mary Foster | *merrillii × sanderiana* | Foster Gdns. | Hawaii |
| 1945 | Maurine Dalton | *bensonii × tricolor* | H. K. Dalton | Hawaii |
| N.A. | Pride of Tjipeganti | *coerulea × insignis* | Tjipeganti | Java[b] |

---

[a] *V.* Herziana and *V.* Gilbert Triboulet should be considered synonymous, as should *V.* Burgeffii and *V.* Tatzeri, because *V. suavis* now is considered to be synonymous with *V. tricolor.*
[b] Not given in *Sander's List of Orchid Hybrids.*

Table 3-2. Strap-leaf *Vanda* hybrids, other than primary, registered up to January 1, 1946 (from *Sander's List of Orchid Hybrids*).

| YEAR | HYBRID NAME | SPECIES PARENTAGE | HYBRIDIZER | PLACE |
|------|-------------|-------------------|------------|-------|
| 1927 | Souvenir de Berthe Jozon | *coerulea* × Gilbert Triboulet | J. Gratiot | France |
| 1933 | Messneri | Burgeffii × *tricolor (suavis)* | Munich Bot. Gdns. | Germany |
| 1933 | Monacensis[a] | Burgeffii × Gilbert Triboulet | Munich Bot. Gdns. | Germany |
| 1933 | Oiseau Bleu | *charlesworthii*[b] × *coerulea* | Vacherot | France |
| 1933 | Schoellhornii[c] | Burgeffii × *tricolor* | Munich Bot. Gdns. | Germany |
| 1933 | Wettsteinii | Burgeffii × *coerulea* | Munich Bot. Gdns. | Germany |
| 1934 | Faustii | Gilbert Triboulet × *luzonica* | Munich Bot. Gdns. | Germany |
| 1940 | Clara Shipman Fisher | *sanderiana* × Tatzeri | H. C. Shipman | Hawaii |
| 1943 | Emily Notley | Mem. T. Iwasaki × *tessellata (roxbh.)* | S. Gillmar | Hawaii |
| 1943 | Frank Scudder | *coerulea* × Mem. T. Iwasaki | S. Gillmar | Hawaii |
| 1943 | Saphir | *coerulea* × Oiseau Bleu | Vach. & Lecoufle | France |
| 1944 | Azur[a] | Oiseau Bleu × Rothschildiana | Vach. & Lecoufle | France |
| 1944 | Puna | *luzonica* × Tatzeri | H. C. Shipman | Hawaii |
| 1945 | Mem. G. Tanaka | *dearei* × Mem. T. Iwasaki | B. Tanaka | Hawaii |

[a] Primary hybrid × primary hybrid.
[b] *V.* × *charlesworthii* can be a natural hybrid of *V. bensonii* × *V. coerulea*.
[c] Same as Messneri.

were made, is not notably better than England's. The British had the largest collections of *Vanda* species in Europe, including fine, well-grown specimens of *V. sanderiana* and *V. coerulea*.

*Vanda sanderiana* did not, at the very start, dominate the making of *Vanda* hybrids as much as might have been expected, given its fine qualities and its dominant role as a parent of both primary and more advanced hybrids after the end of World War II. Only 7 of the 36 hybrids recorded prior to 1946 had *V. sanderiana* as a parent, in contrast with 13 with *V. tricolor* (Plate 9), 12 with *V. coerulea*, and 5 with *V. lamellata* (Plates 4 and 21) and *V. luzonica* (Plate 5).

The Hawaiians did not cut much of a swath prior to the mid-1940s. Hawaii came on strong after that, however, and almost completely took over *Vanda* hybridization until hybridizers in Southeast Asia appeared on the scene in force in the 1960s.

Not until 1927 was a non-primary *Vanda* hybrid recorded. It was *V.* Souvenir de Berthe Jozon (*V. coerulea* × *V.* Gilbert Triboulet), made by Jean Gratiot of France, who had created the first recorded *Vanda* hybrid, *V.* Gilbert Triboulet (*V. coerulea* × *V. tricolor*), registered in 1919.

In the years 1933 and 1934, a small spurt in non-primary hybrid breeding occurred in Germany, together with one new hybrid made in France, mainly using the earlier German hybrid, *Vanda* Burgeffii (*V. sanderiana* × *V. tricolor* var. *suavis*), which combines the same two species as the earlier hybrid, *V.* Tatzeri, and the French hybrid, *V.* Gilbert Triboulet (*V. coerulea* × *V. tricolor*). The popularity of *V. sanderiana* and *V. coerulea* was becoming well established, along with that of *V. tricolor*.

*Vanda luzonica* (Plate 5) was first used as a parent in 1928, when it was crossed with *V. tricolor* (Plate 9) to make *V.* Boschii, a German hybrid originated by the Munich Botanical Gardens. It was used by that institution again in 1934 to make *V.* Faustii (*V.* Gilbert Triboulet × *V. luzonica*). The next big step forward toward the modern *Vanda* hybrid would come when *V. dearei* began to appear in a number of hybrids made in Hawaii after World War II.

The early enchantment with combinations of *Vanda coerulea*, *V. tricolor*, *V. sanderiana*, and *V. luzonica* did not deter breeders from using many other *Vanda* species to make primary crosses, both prior to 1946 and later. Table 3-3 presents a matrix of the primary hybrids recorded up to year-end 1994. By then, 26 species had been used to make 82 different primary hybrids. Most of the listed species are described in Chapter 1.

For the reader's convenience, Table 3-3 can be read either horizontally along the rows or vertically down the columns. This means that each cross is shown twice. For example, a cross of *Vanda sanderiana* and *V. coerulea* is checked in two separate boxes—once in the box where the *V. sanderiana* row intersects the *V. coerulea* column and again where the *V. sanderiana* column intersects the *V. coerulea* row.

A hybrid has the same grex name regardless of which of the parents was the pollen (male) parent and which was the pod (female) parent. The convention in registering a hybrid, however, is to list the pod parent first and the pollen parent second. The simple way to remember this rule is the old maxim: "Ladies before gentlemen." For example, *Vanda* Rothschildiana is registered in *Sander's List* as *V. coerulea* by *V. sanderiana*, but if the cross later was remade in the reverse order, which it has been many times, it would not be registered again under another name; the progeny still would be considered to be *V.* Rothschildiana, even though, in the distribution of genes, it does make a difference as to which plant served as the mother and which as the father. In this latter regard, see Griesbach's (1986) article, *That Reciprocal Cross—Is It a Mule or a Hinny?*, which explains why certain kinds of genes (plastid and mitochondrial genes) are maternally inherited. (One of these, incidentally, influences carotenoid pigmentation, which produces yellow and gold flower colors.)

Table 3-3 does not include natural hybrids. Neither does it enable the reader to learn which of the parents of a registered primary cross was the mother and which was the father. That information can be found in *Sander's List*, which also provides the date the cross was registered and the hybridizer's name. Only the first

Table 3-3. Primary hybrids of strap-leaf *Vanda* species (up to year-end, 1994)

| *Vanda* | *sanderiana* | *tricolor* | *dearei* | *lamellata* | *luzonica* | *coerulea* | *denisoniana* | *merrillii* | *insignis* | *sumatrana* | *tessellata* | *coerulescens* | *cristata* | *bensonii* | *pumila* | *roeblingiana* | *spathulata*[a] | *limbata* | *parviflora*[b] | *brunnea* | *concolor* | *alpina* | *foetida* | *laotica*[c] | *liouvillei* | *stangeana* |
|---|---|---|---|---|---|---|---|---|---|---|---|---|---|---|---|---|---|---|---|---|---|---|---|---|---|---|
| *sanderiana* | | ■ | ■ | ■ | ■ | ■ | ■ | ■ | ■ | ■ | ■ | ■ | ■ | | ■ | ■ | ■ | | ■ | ■ | ■ | | ■ | ■ | | |
| *tricolor* | ■ | | ■ | ■ | ■ | ■ | ■ | ■ | | ■ | ■ | ■ | ■ | ■ | ■ | ■ | | ■ | | | | ■ | | | | ■ |
| *dearei* | ■ | ■ | | ■ | ■ | ■ | ■ | ■ | | ■ | ■ | | | | | ■ | ■ | | ■ | | | | | | ■ | |
| *lamellata* | ■ | ■ | ■ | | ■ | ■ | | | ■ | ■ | ■ | ■ | ■ | ■ | ■ | | | ■ | | | | | | | | |
| *luzonica* | ■ | ■ | ■ | ■ | | ■ | ■ | ■ | ■ | ■ | ■ | | | ■ | | | ■ | | | | ■ | | | | | |
| *coerulea* | ■ | ■ | ■ | ■ | ■ | | | | ■ | | ■ | ■ | | | | | ■ | | | | | | | | | |
| *denisoniana* | ■ | ■ | ■ | | ■ | | | ■ | | | ■ | | | ■ | ■ | | | | | | | | | | | |
| *merrillii* | ■ | ■ | ■ | | ■ | | ■ | | | ■ | | ■ | | | | | | | | ■ | | | | | | |
| *insignis* | ■ | | | ■ | ■ | ■ | | | | ■ | | ■ | ■ | | | | | | | | | | | | | |
| *sumatrana* | ■ | ■ | ■ | ■ | ■ | | | ■ | ■ | | | | | | | | | | | | | | | | | |
| *tessellata* | ■ | ■ | ■ | ■ | ■ | ■ | ■ | | | | | | | | | | | | | | | | | | | |
| *coerulescens* | ■ | ■ | | ■ | | ■ | | ■ | ■ | | | | | | | | | | | | | | | | | |
| *cristata* | ■ | ■ | | ■ | | | | | ■ | | | | | | | | | | ■ | | | | | | | |
| *bensonii* | | ■ | | ■ | ■ | | ■ | | | | | | | | | | | | | | | | | | | |
| *pumila* | ■ | ■ | | ■ | | | ■ | | | | | | | | | | | | | | | | | | | |
| *roeblingiana* | ■ | ■ | ■ | | | | | | | | | | | | | | | ■ | | | | | | | | |
| *spathulata*[a] | ■ | | ■ | | ■ | ■ | | | | | | | | | | | | | | | | | | | | |
| *limbata* | | ■ | | ■ | | | | | | | | | | | | ■ | | | | | | | | | | |
| *parviflora*[b] | ■ | | ■ | | | | | | | | | | ■ | | | | | | | | | | | | | |
| *brunnea* | ■ | | | | | | | ■ | | | | | | | | | | | | | | | | | | |
| *concolor* | ■ | | | | ■ | | | | | | | | | | | | | | | | | | | | | |
| *alpina* | | ■ | | | | | | | | | | | | | | | | | | | | | | | | |
| *foetida* | ■ | | | | | | | | | | | | | | | | | | | | | | | | | |
| *laotica*[c] | ■ | | | | | | | | | | | | | | | | | | | | | | | | | |
| *liouvillei* | | | ■ | | | | | | | | | | | | | | | | | | | | | | | |
| *stangeana* | | ■ | | | | | | | | | | | | | | | | | | | | | | | | |
| TOTAL | 20 | 17 | 13 | 13 | 13 | 9 | 8 | 8 | 7 | 7 | 7 | 6 | 5 | 4 | 4 | 4 | 4 | 3 | 3 | 2 | 2 | 1 | 1 | 1 | 1 | 1 |

[a] Excluded from *Vanda* by Christenson. See Christenson's list of the genus *Vanda*, Appendix B.
[b] *V. parviflora* = *V. testacea*. See Christenson's list.
[c] *V. laotica* = *V. lilacina*. See Christenson's list.

making is recorded; remakes—which may have occurred with parentage reversed—are not included in *Sander's List*. The RHS rules for registration mandate the prior blooming of a plant of the cross.

The diversity of man-made *Vanda* primary hybrids is quite remarkable. For example, *V. sanderiana* has been crossed directly with 20 other species, some of which are rarely heard of. One wonders how the hybridizer was able to acquire the little-known species or their pollen. *Vanda tricolor* (mainly the variety *suavis*) has been mated with 17 other *Vanda* species; *V. dearei*, *V. lamellata*, and *V. luzonica* with 13; *V. coerulea*, somewhat surprisingly in view of its many desirable qualities, with only 9; and *V. denisoniana* and *V. merrillii* each with 8 primary hybrids to its credit.

Of the 26 species listed in Table 3-3, all but 3 of them (*Vanda foetida*, *V. liouvillei*, and *V. pumila*) also have been crossed with one or more hybrids at one time or other. In addition, 2 other strap-leaf *Vanda* species, *V. helvola* and *V. hindsii*, have been used in the creation of *Vanda* non-primary hybrids, although no primary crosses are recorded with any of these as a parent.

Table 3-4 records the number of times each *Vanda* species had served as one of the parents of a *Vanda* hybrid, by year-end 1994. The dominance of *V. sanderiana* and *V. coerulea* is striking. The table tells us something quite significant about what hybridizers were searching for. Although local availability and ease of cultivation undoubtedly have some bearing, the relative use of the individual species presumably is related mainly to two considerations: (1) the apparent aesthetic and horticultural qualities of each species; and (2) the qualities that experience has revealed tend to be transmitted to its immediate offspring and to succeeding generations. If a species has been a parent of a number of successful hybrids, it is reasonable to expect that it is likely to be used by breeders many times.

On the other hand, if hybridizers come to the conclusion that a species seems to make no significant contribution to their breeding program, or that any desirable genes it contains already are incorporated in and well expressed by hybrids presently available, then they have little motivation to continue to use it. This is especially true if a species transmits some undesirable traits along with any desirable ones—which all of them do—because it usually takes several generations to "breed out" the bad traits.

Table 3-4 illustrates the extent to which hybridizers relied on repeated use of species genes as they progressed. Even after allowing for the different lengths of some of the periods, the table shows clearly that hybridizers have made less and less use of species during the last three decades, with the notable exception of *Vanda coerulea*. The table also suggests that few, if any, vastly improved forms of *Vanda* species other than *V. sanderiana* and *V. coerulea* have been discovered either in the wild or through artificial breeding. Intensive line-breeding has produced

Table 3-4. Frequency of use of the *Vanda* species as parents of *Vanda* hybrids[a]
(through year-end 1994).

| VANDA SPECIES USED AS A PARENT | TOTAL THRU 1994 | BEFORE 1946 | 1946 TO 1960 | 1961 TO 1970 | 1971 TO 1980 | 1981 TO 1994 |
|---|---|---|---|---|---|---|
| *sanderiana* | 171 | 6 | 61 | 45 | 36 | 23 |
| *coerulea* | 128 | 13 | 21 | 13 | 32 | 49 |
| *tricolor* (inc. *suavis*) | 39 | 10 | 17 | 4 | — | 8 |
| *dearei* | 33 | 3 | 11 | 13 | 2 | 4 |
| *tessellata* (inc. *roxburghii*) | 32 | 3 | 6 | 9 | 6 | 8 |
| *luzonica* | 29 | 6 | 17 | 4 | — | 2 |
| *denisoniana* | 27 | 1 | 3 | 7 | 9 | 7 |
| *lamellata* | 23 | 5 | 3 | 8 | 1 | 6 |
| *merrillii* | 21 | 1 | 12 | 2 | 4 | 2 |
| *insignis* | 17 | 1 | 8 | 6 | 2 | — |
| *sumatrana* | 9 | — | 5 | 4 | — | — |
| *cristata* | 9 | 1 | 1 | 3 | 2 | 2 |
| *spathulata* | 8 | 2 | 1 | 5 | — | — |
| *coerulescens* | 8 | — | 3 | 4 | — | 1 |
| *bensonii* | 8 | 1 | — | — | 3 | 4 |
| *roeblingiana* | 6 | — | 3 | 1 | — | 2 |
| *concolor* | 5 | — | 5 | — | — | — |
| *lilacina* (incl. *laotica*) | 5 | — | — | 1 | 2 | 2 |
| *limbata* | 4 | — | 1 | 2 | — | 1 |
| *pumila* | 4 | — | 1 | — | 1 | 2 |
| *stangeana* | 4 | — | — | — | — | 4 |
| *alpina* | 3 | — | — | 1 | — | 2 |
| *brunnea* | 3 | — | — | — | 2 | 1 |
| *parviflora* (=*testacea*) | 3 | — | 1 | 2 | — | — |
| *foetida* | 1 | — | — | — | 1 | — |
| *helvola* | 1 | — | — | — | — | 1 |
| *hindsii* | 1 | — | — | 1 | — | — |
| *liouvillei* | 1 | — | — | 1 | — | — |
| TOTAL | 603 | 53 | 180 | 136 | 103 | 131 |

[a] The totals do not record the number of hybrids made with a species as a parent because there is double-counting. For example, V. Rothschildiana (*V. coerulea* × *V. sanderiana*) is recorded once as a hybrid made with *V. coerulea* and again as a hybrid made with *V. sanderiana*.
Hybrids with any non-strap-leaf ancestry are not included in the table.
*Vanda spathulata* is included for historical reasons, although Christenson has placed it in a new genus, *Taprobanea*.
Natural hybrids are not included.

some remarkably beautiful cultivars of both those two species, however, and in the 1980s and early 1990s, *V. coerulea* far outdistanced *V. sanderiana* in popularity as a parent, because of spectacular improvement in the color and shape of its flowers.

During the period 1946–1960, the first postwar period covered by an addendum to *Sander's List of Orchid Hybrids*, *Vanda* hybridizers concentrated on further experimentation with earlier primary crosses, such as *V.* Boschii (*V. luzonica* × *V. tricolor*), *V.* Burgeffii = *V.* Tatzeri (*V. sanderiana* × *V. tricolor*), *V.* Manila (*V. luzonica* × *V. sanderiana*), *V.* Flammerolle (*V. coerulea* × *V. luzonica*), *V.* Gilbert Triboulet (*V. coerulea* × *V. tricolor*), and *V.* Rothschildiana (*V. coerulea* × *V. sanderiana*).

Hybridizers made much use of a 1946 primary hybrid, *Vanda* Ellen Noa (*V. dearei* × *V. sanderiana*). In fact, *V.* Ellen Noa and, to a much lesser extent, *V.* Memoria T. Iwasaki (*V. dearei* × *V. tricolor*) became the primary vehicles for introducing *Vanda dearei* (Plate 2) genes into the lineage of what were to become, down the road, some of the most highly regarded *Vanda* hybrids.

In the 1946–1960 period, a group of four headed the list of *Vanda* species used as parents. Well in the lead was *V. sanderiana*, followed at some distance by *V. coerulea*, *V. tricolor*, and *V. luzonica*. Next in rank were *V. merrillii* and *V. dearei*. Thirteen other *Vanda* species remained in, or were introduced into, breeding programs, and some new primary crosses were registered.

*Vanda merrillii* (Plate 6) deserves special comment. Breeders understandably were excited by its brilliant, lacquered-looking, carmine-red flowers and wished to combine the color, texture, and leathery substance of this species with the larger, flatter flowers available in *V. coerulea*–*V. sanderiana* hybrids. The enchantment with *V. merrillii*, however, was short-lived, because the progeny were disappointing. They lacked large, round, flat flowers and failed to bloom at an early age.

The period 1946–60 was the heyday of Hawaiian breeders of vandas and ascocendas. They were the first to recognize the great potential of *Vanda* hybridizing and to embark systematically on setting the process in motion. They dominated the scene, and they created the gene pools on which today's best hybrids have drawn. Foremost among the Hawaiian pioneers was Herbert Shipman, followed by Oscar Kirsch, Robert Warne, W. W. G. Moir, Richard Tanaka, the Kodama Nursery, and several others. The range of their experimentation and activity was astounding. By their example rather than by words, they set in place the standards still being used today in evaluating the flower quality of vandaceous orchids.

If we begin the period in 1944 instead of 1946, Hawaiian hybridizers registered crosses utilizing a total of 18 different *Vanda* species, in addition to a great many more-advanced hybrids. While a few hybrids were being created elsewhere, they were the exception to the Hawaiian rule.

In the period 1961–1970, the pattern of *Vanda* hybridizing with species underwent several striking changes. The Hawaiians continued to be very active hybridizers, but their interest had graduated to more complex hybrids, and their use of species as direct parents concentrated on *V. sanderiana* and *V. coerulea*.

In the 1960s, the leading role in introducing additional species, and in continued use of some of those previously used, passed to breeders in Singapore and Malaysia, but they did not go on to produce more-advanced hybrids in any significant number. Their main interest was in developing intergeneric hybrids suitable for the burgeoning cut-flower trade.

In the following decade, the Thais preempted the field of contemporary hybridizing of vandas and ascocendas. They adopted the then prevailing standards of the Hawaiian breeders. Refinement, rather than innovation through greater use of species, was the focus. Their goal was large, round, flat, showy flowers, in as large a range of colors and patterns as possible.

Philippine interest in *Vanda* hybridizing never developed, despite the number of *Vanda* species found in that country. Philippine orchid enthusiasts generally import their vandas and ascocendas from Thailand. Political and economic instability are the most likely explanations of the lack of investment in the development of an orchid industry in the Philippines, along with the absence of a leader like Rapee Sagrik, who created widespread public interest in orchids in Thailand.

Overall, use of *Vanda tricolor*, *V. luzonica*, and *V. merrillii* shrank dramatically after 1960 (see Table 3-4). Four species displayed major gains in their use in the 1961–70 period: *V. denisoniana* (Plate 3), *V. lamellata* (Plates 4 and 21), *V. spathulata* (Plate 27), and *V. tessellata* (Plate 8). While the proportionate share of some of the other species also gained, their absolute numbers remained insignificant. *Vanda concolor*, which had been used five times in the 1946–1960 period, subsequently was not used at all.

*Vanda dearei* (Plate 2), which ranked sixth in frequency of use in the 1946–60 period, rose to tie with *V. coerulea* for second place in the 1960s, a notable improvement in its relative position. This is attributable to the appearance of *V. dearei* hybrids made in Singapore and Malaysia, where *V. dearei*, a Borneo species, was readily obtainable, and to the yellow color of its flowers, which had been its major attraction earlier. Another consideration in the frequency of use of *V. dearei* in Singapore probably was the lack of seasonality in the local climate. Hybridizers there tend to prefer species such as *V. dearei* that do not require pronounced seasonality in order to bloom.

In the 1960s and 1970s, the increase in the number of hybrids incorporating *Vanda denisoniana* (Plate 3) was related to the burgeoning interest in orchids in Thailand, where *V. denisoniana* is native, and also to the appealing yellow color of its flowers. There, as in Singapore and Malaysia, hybridizers were using superior forms of locally available material. In Thailand, much credit goes to Rapee Sagarik, who, beginning in the 1960s, actively encouraged popular interest in orchids and their hybridization. That paved the way for the development of a large orchid industry in Thailand, both in potted plants and cut flowers.

Since 1970, the use of *Vanda* species in breeding programs everywhere has shrunk to a point such that, with the notable exception of *V. coerulea*, and possibly *V. sanderiana*, it has become almost an oddity—interesting, but outside the mainstream. One Florida breeder is pursuing a program of hybridization with other *Vanda* species, but he is marching to a drummer different from that of the rest of the hybridizing fraternity, with what success only time will tell.

### Role of Thailand in *Vanda* Hybridizing

Most of the *Vanda* hybrids awarded by the American Orchid Society, and most *Vanda* plants sold anywhere in the world today, are from Thailand.

The strap-leaf *Vanda* industry in that country really began in the late 1960s and early 1970s, using hybrids imported from Hawaii—not expensive, highly selected stud plants but, for the most part, unbloomed seedlings and, especially, community pots of seedlings. (Community pots, usually referred to as "compots," have a cluster of small seedlings in a single flower pot.) At the same time, Thai breeders began to concentrate on producing improved forms of *V. sanderiana* and *V. coerulea* for use in their breeding programs. The ideal climate, inexpensive labor, and (at least "up-country") ready availability of land, all combined to make the costs of production remarkably low. Blooming-size plants can easily be produced within two to two and a half years after the seedlings have been placed in small individual pots or baskets.

The market that commercial growers were targeting was a domestic market—the export market came later, in the second half of the 1970s and early 1980s. Nearly all of those who entered the business had no previous experience with orchids, which had been regarded as a "rich man's hobby," limited to a small number of amateurs. Rapee Sagarik helped establish an active orchid society in the 1960s, and in other ways, such as by a popular TV program, promoted widespread public interest in orchids. By 1970, the interest in orchids, and especially vandas and ascocendas, had taken root and had spread throughout the country. Orchid growing was fashionable and promised to be profitable as well. The latter promise lured many enterprising individuals into the field on a large scale. Few had had any previous experience with horticulture.

During this phase, which, roughly speaking, covered the 1970s, a combination of factors in Thailand paved the way for world supremacy in the production of superior vandas and ascocendas. A burgeoning local demand for orchids, combined with a favorable climate and low labor costs, encouraged the sprouting of a large number of commercial breeders. Also very important, hybridizers could soon see the results of their efforts and could select the best plants for use as parents of the next generation of hybrids. That would not have been possible if most of the plants were exported before blooming.

After the hybridizers made their crosses, they either would grow many of the offspring to maturity or would sell the flasks to other commercial growers to do so. Since most of the nurseries were located within a few hours drive from Bangkok, or in the Chiang Mai area, hybridizers easily could keep track of their competitors' crosses as well as their own. They would purchase the choicest plants from the growers or would arrange to buy or exchange pollen. A great deal of rivalry was evident among local hybridizers; high prices for outstanding plants, and even for their pollen, were common.

Japanese buyers appeared on the scene in the late 1970s and early 1980s. They were interested only in exceptionally good plants, in bloom, and would pay extraordinarily high prices—often several thousand dollars. They were not purchasers of large numbers of plants—only the very best. It is commonly said in Thailand that such plants were destined to serve as gifts for prominent individuals in Japan rather than to be used for future breeding. At any rate, few remain to be seen in Japan. The lure of sales of outstanding plants to Japanese buyers at very high prices induced hybridizers to grow to maturity many plants of promising crosses, instead of selling them earlier.

It is small wonder that the quality of vandas underwent rapid improvement in the 1970s and mid-1980s. The same process occurred with ascocendas. Thai hybridizers grew millions of hybrid plants to maturity and selected the best for their breeding programs. Then they sold the stud plants for high prices and used the finest of the progeny to produce the following generation. Charungraks Devahastin once remarked to me: "Why keep stud plants once they have been used? If a plant fails to produce superior offspring, it isn't worth keeping. And if it does produce some improved progeny, use the best of *them* the next time."

By the late 1980s and early 1990s, however, circumstances had changed appreciably. The export market for orchid plants that began to flourish after the mid-1970s (stimulated in part by admirable displays of Thai vandas and ascocendas at the World Orchid Conference in Frankfurt in 1975) had become the principal market. The domestic market lagged and turned sluggish. While orchid growing still is a popular hobby, it no longer is a craze. Most vandas and ascocendas now are exported before they bloom. The result is that hybridizers are finding it more difficult to monitor and evaluate the outcome of their efforts. The offspring bloom abroad, and the breeder never sees any great quantity of them. Many of the smaller, innovative orchid nurseries have gone out of business, and interest in producing new and improved hybrids has waned considerably. Orchid breeding no longer is regarded as having a high potential for profit. The near-disappearance of Japanese buyers eager to pay extravagantly high prices for premium plants has been a contributing element.

The situation is ameliorated by the fact that both vandas and ascocendas have been developed to such a high state of quality that the potential of the standard gene pools probably has been rather fully exploited. That was not the situation 25 years ago. The challenge today is to stabilize the high quality already achieved. It is unlikely that the current gene pools will produce further major advances in the quality levels already attained with respect to flower number, shape, size, color, and arrangement, as well as age at first blooming and frequency of flowering. Pure yellow or chartreuse varieties of vandas are a possible exception. What is needed is to reduce the considerable amount of deviation from the best specimens. Progress continues to be made in that direction. Fortunately, assessment of progress toward quality stabilization does not require nearly so large a sample for observation as does pursuit of uniquely superior cultivars.

The Thais, unlike the Hawaiians, never have shown much interest in introducing a large number of species into their breeding programs. They did develop much-improved forms of *Vanda sanderiana* and *V. coerulea* and put them to good use. They also have increased the use of *V. denisoniana,* in their search for better yellow-flowering vandas. But their major thrust, from the very beginning, has been to refine and improve the kinds of *Vanda* and *Ascocenda* hybrids the Hawaiians had pioneered, and to adhere to the Hawaiian standards of quality evaluation. In horticulture, as in some other fields, improvement of the known can be as praiseworthy as extending the boundaries of the known.

The Hawaiians probably could not have done what the Thais succeeded in doing, at least not in the same span of time. Their climate and their much higher labor costs, together with the rather small local market, would have precluded production and retention on the scale practiced in Thailand in the 1970s and early 1980s, at the height of the orchid boom there.

## The Winners' Circle

The record of species usage in the making of hybrids tells us where efforts towards innovation and progress were made, but it tells us nothing about the results, other than by inference from the frequency of usage. For determination of success, the best place to turn is the record of flower awards granted by the American Orchid Society. At their monthly judging sessions held throughout the country, and at AOS-approved shows, the judges see hundreds of cultivars of vandas each year—presumably plants that their owners regard as their best. Only a few receive awards. Which species appear in the ancestry of those few, and how far back were they introduced? Experimentation with nearly all of the species listed in Table 3-4 goes back more three decades, so there has been ample time for numerous rounds of selection and parental combinations to have been made and evaluated both for quality and for breeding potential. Any species with much potential should have demonstrated it in its hybrid descendants by now.

When we examine the record of awards during recent years, it comes as a surprise that most of the *Vanda* species tried by hybridizers have left either no legacy of achievement at all, or none worthy of much note. The exceptions are *V. sanderiana*, *V. coerulea*, *V. tricolor*, *V. luzonica*, *V. dearei*, and, to a very limited extent, *V. tessellata*.

A few will argue that the fault lies in the current standards of excellence; that they are too much influenced by the size and shape of the flowers of *Vanda sanderiana* and *V. coerulea*. That may be. The orchid community, like all groups, is greatly influenced by its arbiters of fashion. In this case, the arbiters are the judges of the American Orchid Society. In very large part, however, they mirror the tastes of the larger community of knowledgeable orchid growers; they certainly cannot depart very far from the aesthetic values of their constituency. That constituency prefers *Vanda* flowers that are large, round, flat, and colorful. The same preferences apply to *Ascocenda*.

By the end of the first quarter of 1993, the *Awards Quarterly* of the American Orchid Society had published 148 flower awards to strap-leaf *Vanda* hybrids awarded after January 1, 1984. Ninety-five separate crosses were represented. Three of these crosses produced cultivars that received a First Class Certificate (FCC), 55 others had acquired at least one Award of Merit (AM), and 37 had received at least one Highly Commended Certificate (HCC) but nothing higher.

Eighty-nine of those 95 crosses (94%) have both *Vanda sanderiana* and *V. coerulea* progenitors, and all but 3 of the other 6 grexes have one or the other of these two species in their lineage. All but one of the 58 grexes with cultivars that had won an FCC or an AM had both *V. sanderiana* and *V. coerulea* ancestry. Even for the 37 crosses that produced no plants achieving any award higher than an HCC, 32 have both of these species in their pedigree, and 3 of the remaining 5 have one or the other of *V. sanderiana* and *V. coerulea*. The dominating success of crosses with *V. sanderiana* and *V. coerulea* ancestry is very impressive. Clearly, what they offered is what hybridizers were seeking.

Ninety-three percent of the 95 crosses that produced some award-winners had either *Vanda tricolor* or *V. luzonica* in their background (58% had both), and 69% had *V. dearei*; 48% had all three of them.

Only 5 of the 95 crosses had any *Vanda tessellata* ancestry—3 of these had *V. tessellata* as a very remote ancestor; 2 had *V. tessellata* as a direct parent.

Two crosses had *Vanda denisoniana* in their background. *Vanda cristata*, *V. bensonii* and *V. merrillii* each served as a direct parent of an awarded clone, each of which received an HCC.

Apart from the 10 species enumerated above, *no other* Vanda *species served as a parent or ancestor of any hybrid receiving an AOS flower award in the specified period*. Considering the data in Table 3-4, which record all the individual *Vanda* species that had been used in producing *Vanda* hybrids through the end of 1994 (28 in number), and the frequency of use of each of them, the distribution of awards by ancestral composition is quite astonishing. Many *Vanda* species have been called to serve, but few have served well.

Exploring the matter further, what does the record tell us about the frequency of use of the three *Vanda* species (*V. tricolor*, *V. luzonica*, and *V. dearei*) that, along with *V. sanderiana* and *V. coerulea*, do commonly appear in the ancestry of the best modern *Vanda* hybrids? From the statistics cited about their usage, one might surmise that hybridizers repeatedly used *V. tricolor*, *V. luzonica*, and *V. dearei* as parents in the process of creating the marvels we see nowadays on judging tables and at orchid shows. That is not the situation.

In almost every instance, the only *Vanda* species reintroduced as a parent in later generations have been *V. sanderiana* and *V. coerulea*; one has to go back many generations to find a *V. tricolor*, *V. luzonica*, or *V. dearei* progenitor. There was some small-scale more recent experimentation with *V. denisoniana* as a parent, but, by the early 1990s, it had had little impact on the judging tables. Reintroductions of even *V. sanderiana* were on the decline; while about 20% of the grexes that produced award winners in the period had *V. coerulea* as one of their immediate parents, only 6% of them had a *V. sanderiana* parent. That does not mean that *V. sanderiana* is finished as a stud plant; it does mean, however, that any *V. sanderiana* so used should be truly exceptional.

Such an exception appeared at the end of the 1980s, in the form of progeny of a cross of *Vanda sanderiana* 'Coral Reef', winner of an Award of Merit, and another selected (but unnamed) *V. sanderiana* cultivar. The cross was made by R. F. Orchids, Florida, which has been engaged in an active and successful program of producing fine hybrids. The cultivar *V. sanderiana* 'Robert' was designated the "Best *Vanda* Species" at the 13th World Orchid Conference, held in New Zealand in September 1989. Two other cultivars already have received the highest flower award of the American Orchid Society, a First Class Certificate (FCC). Another of the siblings, *V. sanderiana* 'Orchidgrove', is shown in Plate 23 of this book. Undoubtedly, some of the siblings of the grex will be used successfully as a parent during the coming years.

The genealogy of a *Vanda* grex that received 9 AOS awards between 1984 and 1991, *V. Fuchs Delight* (Plate 32), graphically illustrates some of the observations in this section. Its parents are *V. Kasem's Delight* (Plate 33) and *V. Gordon Dillon* (Plates 34 and 35). Both parents were made in the 1970s by Kasem Boonchoo of Thailand, who has specialized in making large-flowered vandas of excellent quality. Each of these crosses has parented a number of successful offspring, of which *V. Fuchs Delight* is perhaps the most famous. *Vanda* Fuchs Delight, like both its parents, comes in a variety of bright colors—pinks, reds, and blue-violet—and some have a glistening texture as well. Each of the parents provides a good example of the principle that *it is not the frequency of a species' appearance on a hybrid's genealogical chart that is most important for the transmission of its desirable qualities to a hybrid. Rather, it is the number of succeeding generations over which selective breeding has had an opportunity to play its role after the species was introduced.* This same principle applies to other orchid genera as well.

*Vanda* Kasem's Delight, one of the parents of *V.* Fuchs Delight, has *V. tricolor* and *V. luzonica* in its background, in addition to *V. sanderiana* and *V. coerulea*. One has only to look at the flowers of a *V.* Kasem's Delight to conclude that more than *V. sanderiana* and *V. coerulea* genes have made their presence evident. Yet one has to trace the ancestry back to a great-great-great-grandparent of Kasem's Delight before a *V. tricolor* appears, and the same distance back to find a *V. luzonica*. Out of a total of 32 great-great-great-grandparents, only 2 were descended from *V. tricolor* and *V. luzonica*; *V. sanderiana* and *V. coerulea* account for all the rest.

*Vanda* Gordon Dillon, the other parent of *V.* Fuchs Delight, has a similar history. Only three species other than *V. sanderiana* and *V. coerulea* are present in its ancestry, and those three are remote ancestors. *Vanda* Gordon Dillon has one *V. dearei* great-great-great-grandparent, one *V. tricolor* great-great-great-great-grandparent, and one *V. luzonica* great-great-great-great-grandparent.

The genealogy of most of the other award winners in the 1980s and onward is much the same. One has to search a long distance to find any species other than the two dominant ones, *Vanda sanderiana* and *V. coerulea*. By carefully selecting and breeding the best of the progeny of each generation of hybrids over several successive generations after an extremely limited use of a few other *Vanda* species early in the process, hybridizers have been able to produce a rainbow of vibrant colors and unique color patterns (see Plates 29–45).

The record clearly supports a conclusion that one of the most valuable guides for purchasers of unbloomed *Vanda* and *Ascocenda* seedlings is to be found in the pedigree and performance of their parents. A hybrid with *V. tricolor*, *V. luzonica*, and *V. dearei* genes, along with those of *V. sanderiana* and *V. coerulea*, has a somewhat better prospect of producing outstanding progeny than does one with a less complex ancestry—it has a richer gene pool. It is best that any species ancestors other than *V. sanderiana* and *V. coerulea* be remote; otherwise, the undesirable characteristics that most of the other species transmit are all too likely to be evident in the progeny.

# 4

# Criteria of *Vanda* Flower Quality

## THE NEED FOR STANDARDS

The great improvement in *Vanda* and *Ascocenda* flower quality since the 1960s is the result of a sustained process of highly selective breeding. Every plant in a successful hybridizer's collection has had to compete with many others for use as a stud plant, and only the best were chosen. For that competition to be productive, consistent standards and criteria for excellence were necessary to guide the making of selections.

The broad standards that guided Hawaiian breeders when they began large-scale hybridization of vandas in the 1950s remain the same today. Those standards embraced values of the English horticultural establishment, which, in turn, were accepted by the American Orchid Society and throughout the world. Stated very generally, *Vanda* flowers should be large, round, flat, numerous, vibrantly colored, and well arranged on an erect or self-supporting arching inflorescence. Each of these specifications is defined in some detail, so that it can be applied to any individual plant. The terms are relative; the absolute values may change as progress occurs. For example, how large is "large," and how "round" must the petals be to be considered outstanding? Judgments require reference points, and reference points, in turn, are related to what has already been achieved and recorded.

This chapter reviews the current standards in considerable detail, so novices may understand how experts look at *Vanda* inflorescences. Then any novice, should he or she so desire, may apply these standards to his or her own collection. Part of the excitement of growing orchids is that it is a dynamic hobby, not a static one; collections can become outdated as higher levels of achievement are attained. As that late patriarch of the orchid world and past president of the American Orchid Society, Rodney Wilcox Jones, who was constantly renewing his extensive orchid collection, once remarked to me, "One has to keep up with the

71

latest." He was 103 at the time and was still buying young seedlings. He attributed his vitality to his fervent desire not to let his orchid collection "fall behind the times." He lived to 107.

One caveat: Flower standards are not the only, and often not even the principal, consideration that should enter into a decision to keep or replace a plant. At any judging session anywhere, the standards apply only to qualities of a plant that are visible then. It cannot be otherwise. Some very important horticultural traits, however, cannot meet that test: for example, frequency of flowering; uniformity of flower number and quality from one blooming to the next; ease of cultivation; and, not least of all, the lasting quality of the flowers, including resistance to air pollution. A hybridizer would add to this list the ability or prospects of the plant to pass on its good features to its offspring.

Even in the applicable areas, an owner's personal tastes should not be entirely subordinated to the standards applied by judges. That is especially true where new, exploratory lines of breeding are concerned. The standards of national orchid societies provide essential baselines, but they are not the be-all and end-all of orchid evaluation. Nonetheless, it is productive to be acquainted with the general criteria used by the judging community throughout the world—it facilitates meaningful discussion, for one thing—and to have some specific benchmarks as well.

## JUDGING SYSTEMS

The general standards used internationally for judging orchid flower quality are, *broadly* speaking, the same worldwide, both in the characteristics they consider and the respective relative importance of each. The two leading judging systems are those of the United States and Great Britain. These have had great influence on systems used elsewhere. The American Orchid Society employs a rather elaborate and formal point scale to guide its judges in considering possible awards for flower quality. So does the German system, to a lesser degree. The British system does not.

The American Orchid Society has established separate scales for each of the following groups of genera: (1) *Cattleya* (and allied genera), (2) *Cymbidium*, (3) *Paphiopedilum*, (4) *Dendrobium*, (5) *Miltonia* (and *Miltoniopsis*), (6) *Odontoglossum*, (7) *Phalaenopsis*, (8) *Vanda*, (9) the pleurothallid alliance (the genus *Pleurothallis* and related genera such as *Masdevallia* and *Dracula*), and (10) a general scale for plants that do not fit satisfactorily into any of the foregoing sections.

The scoresheet divisions pertaining to each genus are grouped into three broad sections: (1) form (i.e., shape) of the flowers; (2) color of the flowers; and (3) "other characteristics." The total number of possible points for the combined sections is 100, which represents the individual judge's notion of perfection, or the ideal.

With the exception of *Paphiopedilum* and pleurothallids, each of the groups of genera is assigned a total of 30 points for flower form, 30 points for color, and 40 points for all other characteristics combined. (The *Paphiopedilum* scale allocates 40 points to flower form, 40 to color, and 20 to other characteristics, while the pleurothallid category allocates 35 points to flower form, 35 to color, and 30 to other characteristics.)

The two sections that address flower form and flower color are subdivided to evaluate in detail the form and shape of specific parts of the flower (such as its petals and sepals). The weights assigned to these narrower subdivisions are not the same for all of the categories of genera. The "other characteristics" section is subdivided into "size of flower" (always 10 possible points); "substance and texture" (5 to 20 possible points, depending on the genus concerned); "habit and arrangement" of the flowers (always 10 points except for pleurothallids, which are assigned only 5 points for these qualities); and "floriferousness," the number of flowers and buds (10 points). Few highly experienced judges regularly fill out the scoresheet in all its details, nor are they required to do so under the rules, but they are supposed to be guided by the relative weights specified.

The detailed breakdown of the total possible number of points indicates the relative importance the established system assigns to each trait and emphasizes that no single aspect should dominate consideration of the quality of the inflorescence. For example, the flowers of a plant might be regarded as being perfect in shape and color in every respect, but the plant still might not receive an award if the flowers fell far short of perfection in the "other characteristics" category, because the latter category is given a large weight overall. The scoring scale applicable to *Vanda* (which also is used for most intergeneric combinations with *Vanda*) is presented as Appendix C.

A First Class Certificate (FCC) is awarded to a plant if the average number of total points awarded by the individual judges falls between 90 and 100. An Award of Merit (AM) is granted if the average total is between 80 and 89, and a Highly Commended Certificate (HCC) is given if the number is in the 75–79 point range. The range of the individual total scores may not exceed 6 points if any award is to be granted. The purpose of this provision is to assure that a reasonable degree of consensus exists among the judges. But this consensus applies only to the total scores—there are no rules governing acceptable distribution of the subtotals of the detailed categories on the scoresheet, from one judge's scoring to another's. This is a weakness of the system.

The Royal Horticultural Society of Great Britain does not use an explicit scoring system, but the underlying considerations when judging a plant are quite similar. Overall, a plant that would be awarded by the Royal Horticultural Society most probably would be awarded by the American Orchid Society, and vice versa, with the exception that the Royal Horticultural Society bestows only awards of Merit and of First Class, while the American Orchid Society has a third and lower grade of award, the Highly Commended Certificate (HCC). Some American judges regard the latter as a kind of "consolation prize" and would prefer that it not exist; some other American judges hotly dispute that position, however, and believe that the creation of that award was an improvement over the British system. Behind the issue is the implicit question of how elitist the system should be. If the standards for the lowest award are too high, some maintain that too many amateurs will be deterred from exhibiting plants; the HCC award permits more awards to be conferred.

The less structured system used in Great Britain is made possible by the circumstance that there is but one judging committee for the entire country, whereas orchid judging in the United States is conducted at regional judging centers and each center has its own group of judges. There is a need in the United States, therefore, to follow procedures that will promote uniformity of evaluation by the separate groups.

In all judging systems, judges look for improvement over what has been achieved previously. A plant that might well have been highly rated 10 years earlier might now be considered to be of only average quality; hence, the need to update collections constantly, if a high level of quality is to be maintained.

## THE CRITERIA FOR JUDGING

In forming their notion of the ideal, judges must take the ancestral background of a plant into consideration. For example, a hybrid that contains only the three species *Vanda dearei, V. denisoniana,* and *V. pumila* in its ancestry, all of which have short inflorescences with few and rather small flowers, would not be expected to have as many flowers as a *V.* Rothschildiana (*V. coerulea* × *V. sanderiana*), nor would it be expected that the flowers be as large and flat. Plants are supposed to be evaluated *against others of comparable ancestry.*

### Flower Shape

The flowers of *Vanda* and *Ascocenda* hybrids should be round and reasonably flat. Not only should the exterior outline of the individual flowers give an impression of roundness, but so should the outline of each of the petals and sepals.

The relationship of petal width both to petal length and to the overall size of the flower determines whether there is a pleasing balance between the petals and the sepals. Failure to achieve harmonious balance is a shortcoming of the flowers of most *Vanda sanderiana* plants, especially the wild-collected specimens.

I use a somewhat arbitrary rule of thumb that petal width should be equal to at least 90% of petal length. If the petals satisfy that criterion, they will look round. Petal width should be measured on a line perpendicular to the longitudinal axis of the petal, at the point where the petals are widest. On a perfectly shaped flower, that point will be located between one-third and one-half of the distance from the base of the petal. On flowers of mediocre shape, and on most cultivars of *Vanda sanderiana,* the point of intersection is midway or more along the longitudinal axis, rather than closer in. The measurement is taken after flattening the petal.

Another of my criteria for *Vanda* form is that petal width should be equal to at least 45% of the overall span, or natural spread, of the flower. In the AOS awards descriptions, natural spread (unlike the measurement of the individual flower parts) is measured without any manual flattening of the flower, and it is the horizontal measurement unless stated otherwise.

Of the hybrid vandas receiving an American Orchid Society flower-quality award published in the *Awards Quarterly* between the beginning of 1989 and the end of the first quarter of 1993 (20[1]–24[1]), 72% met both of these numerical criteria for petal width (i.e., petal width equal to at least 90% of petal length and 45% of natural spread). That compares with 49% of the awarded hybrid vandas published in the *Awards Quarterly* in the three years 1980–82, and 19% for the six years 1969–74. Clearly, the shape of awarded vandas improved demonstrably as hybridizers persisted in their efforts to improve flower shape. Improvement in flower shape ranked first in the goals of Thai hybridizers of *Vanda* hybrids.

Absolute size of the petals also is a consideration, so long as the proportions of the flower segments are harmonious. A *Vanda* flower with a petal width of 4.5 cm (1.75 in) is considered to be of quite good size. Adding this measurement as a third standard of petal width, we find that 70% of the awarded vandas met *all three of the criteria* in the 1989–93 period, while, in the 1980–82 and 1969–74 periods, the ratios were 41% and 12%, respectively. If we push the petal width criterion up to 5 cm (2 in), then 55% met all the tests in the 1989–93 period, as compared with 25% and 4%, respectively, in the two earlier periods. Better shape has not been achieved at the expense of size.

In addition to petal width, other traits affect the desirability of a *Vanda* flower's shape. The length of the dorsal sepal should not be much longer than the length of the petals; the dorsal sepal should not stick up above the tops of the petals like a sore thumb. Neither should it be much longer than the lateral sepals. The petals should nearly touch each other where they overlap the dorsal sepal.

Ideally, the bottom edge of the petals should touch, or slightly pass, an imagined horizontal line drawn at the same elevation as the base of the side lobes of the lip. That produces an appearance of balance, on the one hand, between the petals and the dorsal sepal and, on the other hand, between these and the lateral sepals below. On flowers of *Vanda sanderiana,* and also on some of the hybrids closely resembling it, the bottom of the petals touches a higher horizontal line drawn through the underpart of the *column* of the flower, rather than through the base of the side lobes, with the result that the upper and lower portions of the flower are not pleasingly balanced.

Preferably, one of the lateral sepals should overlap the other, and the pattern should be uniform for all the flowers on the inflorescence. If, for example, the left lateral sepal overlaps the right one on one of the flowers, then that pattern should apply to all the other flowers as well. A little manipulation by the grower can correct an occasional deviation on an inflorescence, in most instances.

Sometimes the lateral sepals do not overlap, despite being amply broad, because the inner edge of both sepals is rolled back, and the two rolled portions touch each other, instead of one sepal passing over the other. How much of a fault this is will depend partly on the breadth of the lateral sepals and on the amount of backward roll. If the lateral sepals are very wide and a sizable portion is rolled back, and thus not visible, the effect will be aesthetically unpleasing. On the other hand, if the sepal width is modest, the sepals would not have overlapped much even if their inner margins were flat; in that case, a small amount of recurving may make the two rolled portions touch each other in a way that, while not ideal, is not a major fault. The inner margins of the lateral sepals of *Vanda* hybrids with a high proportion of *V. coerulea* ancestry almost always roll back slightly; the sepals touch, rather than overlap.

On some *Vanda* flowers, the lateral sepal, or both lateral sepals, will have an extremely narrow *forward*-curving roll or flat crease on its inner margin where the lateral sepals overlap or touch each other. Generally, it is no more than 1.5 mm (.06 in) wide. Oftentimes it extends along the bottommost edge of a lateral sepal. This trait is found on some cultivars of *Vanda sanderiana* (for example, see Plates 251 and 255 in Volume II of Valmayor's book [1984]). It is reasonable to assume that, when this trait is found in hybrids, it is inherited from that species.

If the rolling or creasing is scarcely noticeable, it is a very minor blemish. In some plants, however, it is unacceptably obtrusive. Thai growers seem to be less disturbed by this oddity than are Americans and Europeans. The flaw appears on some plants of the famous hybrid *Vanda* Kasem's Delight, although usually to a very minor degree; for example, see Plate 33, *V.* Kasem's Delight 'Krachai'. The other fine qualities of this cultivar more than compensate for the very faint creasing of the inner edges of the lateral sepals, and the plant was awarded an HCC by the American Orchid Society.

The petals and sepals should be quite flat over their full length. Some flowers have petals or dorsal sepals that are recurved, or pinched looking, at their base where they join the column, even though the outer halves of the petals or dorsal sepals may be only slightly convex. This is a fault frequently seen in hybrids with a *Vanda coerulea* parent (see Plate 46), because the petals and dorsal sepals of that species tend to be clublike or clawed, with a rather narrow, stalklike base. The flowers of the *Ascocentrum* species in the ancestry of today's *Ascocenda* hybrids also sometimes are somewhat club-shaped; so they, too, have contributed this trait to some ascocendas. In breeding programs, *V. sanderiana* is the species that best counteracts the problem of pinching at the base of the petals and sepals, especially if one of the greatly improved specimens of *V. sanderiana* is used as the *Vanda* parent.

Some plants have sepals and petals whose broadest surfaces are convex, while others have a pronounced concavity, or cupping, of those surfaces. Cupping, combined with a forward tilt at the top, frequently occurs in the dorsal sepal, creating a hooded effect over the column—not a desirable feature,

Cupping is a condition that gives rise to a fair amount of discussion at orchid judgings. Perfect flatness appears to be the standard implied in the American Orchid Society *Handbook on Judging and Exhibition* (1991). Yet, to my mind, a *Vanda* flower that is *completely* flat (a characteristic seen once in a great while) tends to look as if it had been cut out of a sheet of cardboard or plastic. Personally, I find a very small amount of *overall* concaveness pleasing, but anything more than that is not, and excessive cupping is a common fault of vandas and ascocendas. Cupping, if at all pronounced, obscures the view of individual flowers when an inflorescence is viewed from different angles. It also reduces the apparent size of the flowers.

Since the measurements published in the *Awards Quarterly* record the natural spread of the flower unflattened, while the petals and the other flower segments are measured after being extended to their full length, the ratio of petal length to recorded natural spread is an indicator of the degree of cupping. The higher the ratio, the more the cupping, because, as cupping increases, the measured natural spread is diminished and the ratio of (manually extended) petal length to natural spread is raised.

The maturity and health of a plant affect the degree of flatness of its flowers, as well as their size. Accordingly, if a *Vanda* or *Ascocenda* has cupped flowers, the problem may lie, at least in part, in culture or immaturity, and not necessarily in heredity. A plant should not be discarded too quickly if cupping is its main fault; judgment should be passed only after at least two successive bloomings on a strong, healthy plant with an abundance of active roots and turgid leaves.

In short, the shape of the individual flower segments of vandas and ascocendas often is either too convex or too concave. The desired impression is essentially one of flatness, both overall and for each of the petals and sepals individually, to create an ensemble of beauty. The worst of all situations is one where an individual flower simultaneously displays areas of convexity and concavity, presenting a bumpy surface.

The outline of a *Vanda* or *Ascocenda* flower should have simple and free-flowing lines. The edges of the petals and sepals should be clean cut, not ruffled or scalloped. When the latter conditions occur, they probably are a legacy of the non-V. *sanderiana* ancestry of the hybrid. While some viewers, even an occasional orchid judge, regard some ruffling to be rather attractive, it does not take much of it to constitute a fault. A glance at the picture of *Vanda* Michael Coronado (Plate 41) illustrates how much the *absence* of such a trait contributes to the beauty of a flower.

The lip of a *Vanda,* as succinctly phrased in the American Orchid Society *Handbook on Judging and Exhibition* (1991, 31), "in size and shape should be harmonious with the rest of the flower, in accordance with the ancestral species." The same applies to ascocendas and other intergeneric combinations with *Vanda.* The lip is not a prominent feature of most *Vanda* hybrids, and fewer points are assigned to lip size than are assigned to that measurement in the case of the other genera listed on the scoresheet.

### Size of Flowers

The expected average flower size of the offspring of a cross of two orchids is the geometric mean of the flower size of the parents, not the arithmetic mean. For example, if one parent has a natural spread of 12 and the other has a natural spread of 3, the expectancy for the progeny will be equal to the square root of ($12 \times 3$), which is 6, rather than the quotient of ($[12 + 3] \div 2$), which is 7.5.

A good simple reference point for flower size is the arithmetic mean of the natural spread of awarded plants. The descriptions of awards published in the *Awards Quarterly* of the American Orchid Society from the beginning of the first quarter of 1989 through the first quarter of 1993 (20[1]–24[1]) cover 69 strap-leaf hybrid vandas that received awards for flower quality. (Awards also are given to growers for culture; these are not included here.) The average natural spread of the flowers of these 69 plants was 10.6 cm (4.2 in). That is an insignificant increase over the corresponding figure for awarded plants in the 1969–74 period, which was 10.4 cm (4.1 in).

The average figures for all awarded plants do not tell the whole story, however, because there was a significant increase in the average size of the *best* plants. Thirty-eight of the 69 plants received an Award of Merit (AM), requiring a score in the range of 80–89 points; one additional plant received a First Class Certificate (FCC), requiring a score of at least 90 points. The average natural spread of the flowers of those 39 plants was 11.2 cm (4.4 in), up from 10.4 cm (4.1 in) in the 1969–74 period—a substantial improvement of 0.8 cm (0.3 in). The median natural spread, theoretically an even better measure than the arithmetic average because it is less sensitive to extremely large or extremely small figures, was the same as the arithmetic average—11.2 cm (4.4 in).

For the 30 plants receiving an HCC, an award requiring a score in the 75–79 point range, the average measurement actually declined from 10.2 cm (4 in) in the earlier period to 9.8 cm (3.9 in) in the later period. The decline was a result of three "novelty-type" crosses that received HCC awards; each of them had one of the smaller-flowering *Vanda* species as a parent, and that diminished the size of the flowers of the offspring. If those three plants are excluded from the tally, the average natural spread for the more typical plants receiving an HCC is raised to 10.3 cm (4 in), a slight improvement over the 10.2 cm (4 in) average size of the earlier period.

The increase in the average flower size of the upper echelon of awarded plants (those receiving better than an HCC award) is a noteworthy advance. Hybridizers in the late 1980s and early 1990s began to give much more attention to flower size, having achieved the colors and color patterns they were seeking earlier. Until the late 1980s, most hybridizers in Thailand believed there was a conflict between color intensity and flower size, and they gave greater priority to the former. Now they know that is not so; consequently, they are placing more emphasis on flower size when they choose plants to use as breeding stock.

One leading Thai hybridizer, Kasem Boonchoo, has been producing some *Vanda* hybrids with remarkably large and flat flowers by using a tetraploid form of *V.* Kasem's Delight (as well as some other tetraploid vandas) as a parent. The results often are spectacular.

The comparative data support three conclusions about the flower size of *Vanda* hybrids: (1) the higher the award, the larger the average size of the flowers, as one would expect; (2) the average size of the flowers of awarded plants did not increase until about 1989, unlike the earlier progress discernible in petal width and overall flower shape; and (3) the flowers of a *Vanda* cannot be considered of exceptional size unless they measure close to 11 centimeters (4.3 in) across, or preferably a bit more.

As in the case of flower shape, flower size depends in considerable measure on the strength and maturity of the plant. One legendary grower, Marilyn Mirro, of Long Island, New York, consistently produces *Vanda* flowers with larger size than anyone else in the region, and of good shape and substance. I am convinced she does not have only plants that are exceptional genetically; rather, she is a superb grower, and no one else approaches her in this respect. Her secret: *attention, attention, attention,* continually and consistently (Mirro, 690–695). But be forewarned: few persons will be prepared to dedicate themselves to their plants to the same extent she does.

## Number of Flowers

The rule governing genetic expectancy of the average number of flowers likely to be borne on an inflorescence of the offspring of a cross is the geometric average of the numbers typically borne by the two parents, not their arithmetic average, just as is true of the expected size of the flowers. In the judging of individual plants, however, the most practical measure is how well a plant's number of blooms compares with the median number of flowers of the plants chosen as relevant for comparison (e.g., those of the same type that have received awards).

In flower awards published in the *Awards Quarterly* from the beginning of 1989 through the end of the first quarter of 1993, the median number of flowers on a hybrid *Vanda* inflorescence was 13 for AMs and 12 for HCCs—the same as in the 1969–74 period. Table 4-1 shows a breakdown by AM/FCC and HCC awards. Note that the changes in the distribution from the earlier to the later period are not large, except that the number of flowers on plants receiving higher awards (AMs and FCCs) tends to cluster in the 10–17 bracket more than was the case in the earlier period. Based on the *Awards Quarterly* data, 12 flowers per inflorescence may be regarded as a reasonable standard of excellence.

A comparative analysis of the data discloses that there is no inverse relationship between the number of flowers found on awarded hybrid *Vanda* inflorescences and the size of the flowers; thus, flower size need not be sacrificed in order to obtain more flowers. That is not true of ascocendas. There, the greater the number of

Table 4-1. Floriferousness of awarded vandas.

| NUMBER OF FLOWERS AND BUDS PER INFLORESCENCE | TOTAL | | AMS AND FCCS | | HCCS | |
|---|---|---|---|---|---|---|
| | 1969–74 | 1989(1)–93(1) | 1969–74 | 1989(1)–93(1) | 1969–74 | 1989(1)–93(1) |
| | | | (PERCENT OF TOTAL) | | | |
| 5–9 | 12% | 12% | 6% | 5% | 20% | 20% |
| 10–13 | 45% | 46% | 45% | 46% | 46% | 47% |
| 14–17 | 31% | 38% | 35% | 46%[a] | 26% | 27% |
| >17 | 12% | 4% | 14% | 3% | 9% | 7% |

[a] Includes 1 FCC award.
Source: Data calculated from awards descriptions in the *Awards Quarterly* of the American Orchid Society.

flowers, the smaller the average size of the flower, generally, and, as the size of *Ascocenda* blossoms increases to near *Vanda* proportions, the number of flowers is reduced to the smaller quantity characteristic of vandas. In ascocendas, the number of flowers is greatly influenced by the proportionate representation of *Ascocentrum* in the ancestry of the hybrid.

## Flower Color, Pattern, and Texture

One of the most appealing features of vandas and ascocendas is their striking color, but the feature is one of the most difficult to describe because of the tremendous diversity of hue, intensity, and patterns of color. The same is true of other intergeneric combinations with *Vanda*. Several pigments combine to create a flower's color or colors. Most of these pigments are affected by the pH of pigment-containing cells of the flowers; the pH of those cells, in turn, is controlled by genes. Some genes alter the hue and intensity of the pigments, and some control the location or distribution of the pigments, resulting in an incredible range of possible combinations.

Color is as complex a phenomenon visually as it is chemically. What seems to be a solid red or violet color, for example, under magnification often is found to consist of a multitude of minute red or purple dots densely spread over a background of, say, creamy white. The color of the background makes a great difference in how flower color appears to the viewer. It is especially relevant to the brightness of color, and its depth or intensity. The ambient lighting also makes a great difference in how colors are perceived.

The surface texture of the sepals and petals also has a bearing on color. The best surface texture is one where the pigments appear to be crystallized; that gives the colors a luminous quality that is very appealing, even at a considerable distance. In the 1970s, sparkling texture was seen in the best of the early *Ascocenda* hybrids, and it was one of their great attractions, but it was not often seen in *Vanda* hybrids. Much progress in that direction, however, has been made by *Vanda* hybridizers since then. Now, some of the best vandas can nearly match the best ascocendas in this respect, even though that quality is not yet the norm.

The distribution of pigments is a significant factor in the perception of color. For example, in some flowers, the pigmentation runs deeply through the sepals and petals; in some others, it is merely on the surface (as with flowers that are strongly colored on their front but whitish on their rear). We know little about the process by which pigments are distributed, and even less about how to manipulate the process in the course of hybridization. Trial and error seem to be the only approach, although using a tetraploid parent often helps to improve depth of color, because of the heavier dosage of color-controlling genes.

Over the past decade, *Vanda* and *Ascocenda* colors have become much deeper and more vivid, to such an extent that vandas and ascocendas with dull or "muddy" flowers, or faded-looking ones, hardly deserve to be kept in any collection, no matter what the other merits of the plants. The surface texture of a good modern *Vanda* or *Ascocenda* should be crystalline under favorable lighting conditions, and any tessellation, venation, bars, stripes, blotches, and other markings should be sharply delineated. Solid colors should extend undiminished to their boundaries—they should not become paler at the periphery of the petals and sepals.

The combination of colors and markings should be visually pleasing. For example, a *Vanda* with greenish yellow flowers should leave the viewer with a cool, refreshing sensation, like lemon or lime sherbet, and not remind one of bile. A good *Vanda coerulea* hybrid should suggest royal-purple velvet, not blue denim that has been laundered too often. A good *Ascocenda* should have a sparkling quality when viewed in bright daylight. In short, what is sought is vibrancy.

As mentioned in Chapter 3, hybrids distantly descended from *Vanda tricolor*, *V. luzonica*, and *V. dearei*, and having *V. sanderiana* and *V. coerulea* among their more recent ancestors, have the greatest likelihood of producing progeny with outstanding color. Chapter 5 discusses the subject with specific reference to ascocentrums and ascocendas.

## Flower Substance

Flower substance refers to the thickness and firmness of the tissue of the flowers. Substance generally has not been a problem with *Vanda* and *Ascocenda* hybrids and has not required special attention from hybridizers. Some of the triploids appearing on the market have extraordinary substance. At the same time, triploidy in vandas sometimes is accompanied by a slightly elevated ridge, or rib, beginning at the base of each petal and proceeding, in a tapered form, along the centerline of the petal for a distance of about 2–2.5 cm (0.8–1 in). This rib thickens and strengthens the axis of the petal and makes it look reinforced structurally, which indeed is its function, as is the case with the slightly elevated and thickened rib seen on many large white *Phalaenopsis* petals.

This feature may be regarded as a mark of sturdiness and strength, contributing to the ability of the petals to support themselves in a flat, erect position, or it may be considered a scar on a pretty face. It is neither a plus nor a minus as long as it is not so gross in elevation and thickness that it is distracting. As larger flowers are produced by triploidy, this feature can be expected to occur more often; larger flowers require greater substance and strength to stiffen their petals.

## The Flower Stalk

Everyone wants *Vanda* flower stalks to be erect or gracefully arching. They also should be long enough to bear a goodly number of large flowers held well above the leaves and not bunched together. Those desirable qualities cannot be achieved, however, without putting quite a strain on the flower stalk. The weight of the flowers may cause it to kink or break unless it is very strong (i.e., thick and hard) or is supported in some fashion. Breaking is most likely to occur after the plant has been drenched, because then the additional weight of the wet flowers is more than a long weak stalk can bear. Using a tie or stake to hold up a *Vanda* inflorescence is not well regarded by discriminating observers, although it is tolerated by some judges.

Weakness of the flower stalk is a genetic trait for the most part, but excessive use of nitrogen in the feeding regimen, and too little light, can magnify the expression of a genetic tendency. *Vanda coerulea* plants characteristically have long, rather thin, flower stalks, and tend to impart this trait to their hybrids. *Vanda sanderiana* plants, on the other hand, have exceptionally strong, thick flower stalks of moderate length. Unfortunately, until the latter part of the 1980s, hybridizers did not pay enough attention to the stoutness or stiffness of the flower stalks; they were giving greater priority to other desired qualities. Although they succeeded in their other objectives, they often found that the flower stalk was weak, especially with *Vanda* hybrids with substantial amounts of recent *V. coerulea* ancestry. Since then, hybridizers—at least the best breeders—have been giving more emphasis to correcting this problem through their selection of stud plants.

When choosing among plants in bloom, it is advisable to avoid those with spindly flower stalks unless they have enough outstanding qualities to make one somewhat forgiving of this weakness, but when you encounter it, be prepared to stake the inflorescences or support them in some other manner. Those with crooked or twisted stalks almost always should be passed by. Most judges regard this condition as a deformity. The last fault sometimes is a legacy of non-*Vanda sanderiana* and non-*V. coerulea* ancestry.

## Pedicels

Pedicels are the stalks of the individual flowers; they are composed of the ovary plus the stem below the ovary to the point of attachment to the flower stalk. The pedicels of *Vanda* flowers have spiraling ridges on their surface because the flowers are resupinate; they emerge as buds in an upside-down position and then gradually twist 180° as they mature and open to what is considered the correct posture, with the dorsal sepal heading upward rather than downward from the column. The twisting causes the spiraling in the ridges that run along the length of the pedicel, including the ovary. Resupination facilitates access to the nectary by pollinators, by making the lip a landing platform.

Pedicels should emanate from the flower stalk horizontally or, even better, with a slight upward tilt. They should be stiff enough to support the flowers and to maintain the angle even when the flowers are wet and heavy. Their length and spacing should be sufficient to prevent crowding of the flowers but should not be so pronounced as to create an excessively open effect.

When pedicels are too short, the flowers are pulled toward the stalk to such an extent that they overlap and crowd one another, especially if the flowers are large. This "bunchy" effect is a common problem with inflorescences of *Vanda sanderiana,* and the trait often is transmitted to its hybrids.

Excessively long pedicels (see Plate 47) create two kinds of aesthetic problems: poor flower posture and poor flower arrangement. The longer the pedicels, the harder for them to support the flowers in an upright position, especially if the pedicels are thin. The flowers may tilt downward, making it difficult for viewers to enjoy their beauty at eye level. As breeders strive for larger, heavier, and more numerous flowers, therefore, they also have to concentrate on stronger flower stalks and pedicels. *Vanda sanderiana* flowers have stout stalks and pedicels, and the species transmits them to its progeny. It is the only commonly used *Vanda* species that is outstanding in this regard.

Excessively long pedicels also adversely affect flower arrangement (see Plate 48). The flowers look "spacey," to use a term popular with Thai connoisseurs. That simply means that the individual flowers have too much space between them. The problem of spaciness is considerably aggravated if the distances between the bases of the pedicels, where they join the stalk, are greater than normal, and also if the flowers are small in size.

### Arrangement of the Flowers

Criteria of flower arrangement go beyond pedicel length and angle. Flowers should encircle the stalk uniformly and be well presented from any vantage point. The faces of the flowers should be held in a vertical position and not hang as if in shame; that is why the pedicels should emanate from the flower stalk horizontally or, even better, with a slightly upward inclination. The inflorescence should be erect, or nearly so, except that very long inflorescences may be gracefully arched. The flowers should begin far enough up the flower stalk so that they are not hidden or crowded by nearby leaves.

For the flowers to be both large and well arranged, the flower stalk must be long enough to prevent excessive vertical overlapping. Vertical (or horizontal) overlapping prevents the viewer from seeing and enjoying the beauty of each blossom individually.

Just as overlapping and crowding are distracting, so is higgledy-piggledy arrangement. The dorsal sepal of each flower should be perfectly vertical. It is disturbing to have flowers cocked at other angles, which sometimes occurs; worst of all are flowers cocked at a variety of angles.

Inflorescences with the flowers bunched together near the top, like a lollipop, are unattractive. This was a common fault of early hybrids with much *Vanda sanderiana* ancestry in their immediate background. The longer inflorescences of *V. coerulea,* and perhaps also *V. luzonica,* helped to spread out the flowers. *V. dearei,* with its short racemes, aggravates the tendency of *V. sanderiana* toward bunchiness; the legacy still is seen in many of today's yellow-flowering vandas.

Acceptable arrangement also is related to whether the pedicels of the flowers are attached to the main stalk in a symmetrical pattern. If there is uniform spacing of the pedicels, both horizontally and vertically, the flowers can be somewhat closer together without appearing to be crowded, or they can be somewhat farther apart without seeming spacey. But if there are pronounced variations in spacing, flowers on one part of the inflorescence will appear crowded while those on another section will look spacey—the worst of both worlds. Uneven spacing of individual flowers is one of the first details to catch the eye of a perceptive observer.

Acceptable arrangement, oddly enough, is not a matter simply of fixed geometric relationships. Rather, it is influenced by the imagination—by what the mind imagines as well as by what the eyes actually see. For example, if the flowers on an inflorescence are a solid color, they can be spaced somewhat more closely together and still give a visually pleasing impression. On the other hand, multicolored and highly patterned flowers require somewhat greater spacing. Why? Because it is much harder for the eye to visualize the obscured portions that are an important part of the character of such flowers. With flowers of a single color, the eye does not have to distinguish a variety of colors and their positioning, so it is less essential to see exactly where one flower ends and another begins. Of course, any overlapping cannot be extreme—then the eye would envisage just a blob of color rather than an arrangement of flowers.

Acceptable spacing requires either that the eye can readily discern the entirety of each individual flower without being distracted by excessive crowding or gaps, or that the mind's eye can imagine that it does, because it can fill in the obscured positions or vacant spaces without straining.

## Branching

Occasionally, branching occurs on the inflorescences of vandas and ascocendas. Branched inflorescences generally have more flowers than the normal unbranched ones, so that would seem to be an advantage. Branching also tends to cause crowding, however, and it prevents the viewer from seeing some of the individual flowers. Branching is not an advantage in vandas or ascocendas; it is a distraction, especially because it seldom is symmetrical.

## Novelty Crosses

Criteria are well established for evaluating flower quality of standard *Vanda* and *Ascocenda* hybrids, but few specific guidelines exist for the novelty type of hybrids. This presents a challenge with respect to some of the delightful intergeneric combinations that are being made, especially with *Aerides* and *Rhynchostylis,* and it would be helpful if some practical guidelines were to be developed for comparative and evaluative purposes. These would have to be primarily nonnumerical criteria.

The cardinal rule for novelty crosses is that the progeny should be better than both parents in one or more important respects, and not markedly inferior in any other regard. The progeny should be innovative and distinctive in appearance, obviously in a way that is attractive. If a plant from such a cross meets those tests, it then becomes largely a matter of the judges' personal tastes and expectations whether the plant receives a high score. Particularly important is what the individual judges imagine that the cross should be capable of producing, given their knowledge of other crosses made with similar parents.

Novelty hybrids can have truly charming qualities that make them an interesting addition to one's collection. Two older examples that come to mind are *Vanda* Ben Berliner (*V. cristata* × *V. coerulescens*), registered in 1969, and *V.* Paki (*V. cristata* × *V. tricolor*), registered in 1944. Both of these hybrids derive much of their appeal from their prominent, highly-colored flower lips, which clearly show the influence of the *V. cristata* parent (see Plate 20).

More recently, some exciting *Christieara* (*Aerides* × *Ascocentrum* × *Vanda*) crosses have appeared on the scene, as well as some *Vascostylis* (*Ascocentrum* × *Rhynchostylis* × *Vanda*) and *Kagawara* (*Ascocentrum* × *Renanthera* × *Vanda*) hybrids. Other intergeneric combinations with *Vanda* are appearing more often. Some of these innovations have eye-catching lip structures, striking new combinations and blending of flower colors, and other traits that attract favorable attention. Except for hybrids developed specifically for the cut-flower trade, these developments are not much beyond their infancy. Given the genetic diversity and richness of the potential combinations, it will be some time before their potential for hobby-growers can be fully appreciated and thoroughly evaluated. Fortunately, exploration of these avenues of breeding is occurring in many places: in Florida, Hawaii, Malaysia, Singapore, and Thailand. That should help to increase the availability of new and exciting crosses. Chapter 6 explores many of the intergeneric crosses with *Vanda* in much greater detail.

# 5

# Asocentrums and Ascocendas

## ORIGIN AND COMPOSITION OF THE GENUS *ASCOCENTRUM*

The genus *Ascocentrum* Schltr. was established by the German botanist Rudolf Schlechter (1913, 975). Prior to that, members of the genus were regarded as belonging in a broadly defined *Saccolabium*. The name *Ascocentrum* is derived from the large spur that hangs from the lip of the flower.

Ascocentrums are small, compact epiphytes that somewhat resemble vandas in their growth habit and general appearance. They come from roughly the same regions of the world. There are two vegetative types: one has strap-shaped leaves; the other has semi-terete leaves, deeply furrowed on the top surface.

The flowers of *Ascocentrum* species are small and look rather like miniature vandas, but differ in three important respects: the sepals and petals are nearly equal in size; the flowers are more brightly colored; and they are far more numerous and tightly arranged on their flower stalks than are those of vandas.

Ascocentrums are delightful species in their own right, but their principal contribution to horticulture has been in the creation of the man-made genus *Ascocenda*, a combination of *Ascocentrum* and *Vanda* that often displays the best characteristics of the species of both genera.

As in the case of the genus *Vanda*, placement in the genus *Ascocentrum* has been subject to some confusion and substantial revision over the years. Until very recently, the genus might be said to contain 9 or 10 species. Christenson (1986b, 105–107, and 1991, personal communication) includes the following: *Ascocentrum ampullaceum* (Roxb.) Schltr.; *Asctm. aurantiacum* Schltr.; *Asctm. aureum* J. J. Smith; *Asctm. curvifolium* (Lindl.) Schltr.; *Asctm. himalaicum* (Deb. Sengupta & Malick) E. A. Christ.; *Asctm. miniatum* (Lindl.) Schltr.; *Asctm. pumilum* (Hay.) Schltr.; *Asctm. rubrum* (Lindl.) Seidenf. (Seidenfaden 1988, 309–318); and *Asctm.*

*semiteretifolium* Seidenf.; as well as the somewhat aberrant *Asctm. pusillum* Averyanov, from Thailand and Vietnam.

In addition, as of the end of the first quarter of 1993, four new *Ascocentrum* species from Borneo, the Philippines, and Indochina had just been, or were in process of being, published by botanists. The four are: (1) *Asctm. insularum* E. A. Christ. (1992), a new species from East Borneo that has deep brownish red flowers, and differs from *Asctm. miniatum* by its smaller flowers and its sharply truncate concave labellum midlobe; (2) *Asctm. aurantiacum* ssp. *philippinensis* E. A. Christ. (1992), a subspecies from the Philippines—previous Philippine specimens ascribed to *Asctm. miniatum* were so ascribed in error and are referable to *Asctm. aurantiacum* Schltr.; (3) *Asctm. christensonianum* Haager, a new species from Vietnam (Haager 1993, 39–40) that resembles a light-colored *Asctm. ampullaceum;* and (4) *Asctm. garayi* E. A. Christ. (Christenson 1992), a new name for Indochinese plants that had usually been called *Asctm. miniatum.*

In horticulture, especially in nursery catalogs, references to a Borneo species called *Ascocentrum hendersonianum* are encountered, and the American Orchid Society has given awards to the species under this name. Traditionally, however, taxonomists have included the species in the genus *Saccolabrium,* not in *Ascocentrum.* In 1986, Christenson (1986c) placed the species in a new monotypic genus named *Dyakia,* derived from the Malay word 'Dyak,' which refers to the indigenous inhabitants of Borneo. The name *Dyakia hendersoniana* now is widely accepted by botanists but has not appeared much in horticulture. The plant is a true miniature. Flower color is variable, from magenta to coral, in various shades; the lip and spur are white. Flower size is 0.6 cm (0.25 in). No hybrids have been registered with it as a parent.

Holttum (1972, 735) transferred *Saccolabium micranthum* to *Ascocentrum* in 1947, but in 1960 he moved it again, this time to a new genus he named *Smitinandia,* in honor of Tem Smitinand, a Thai scientist.

*Ascocentrum rubrum* (named *Saccolabium rubrum* when it was first described in 1833 by the pioneering orchidologist, John Lindley) raises some interesting questions with respect to the record of hybridizing with *Asctm. curvifolium.* Generally, *Asctm. rubrum* has been considered conspecific with *Asctm. curvifolium,* and very rarely is even mentioned. It differs from *Asctm. curvifolium,* however, in that the midlobe of the latter species is truncate retuse, while that of *Asctm. rubrum* is obtuse. (See Bechtel, Cribb & Launert [1981, 431; 1992, 571] for diagrams of the two shapes.) Flowers are variable red and measure 2.2 cm (0.87 in). Seidenfaden (1988, 313) not only gives *Asctm. rubrum* species status, he also states:

> I am inclined to believe that many of the very red flowers in photographs called *Ascocentrum curvifolium* could rather be *A. rubrum,* e.g., I suspect that the picture by Kamemoto and Sagarik (1975, 172) represents *A. rubrum.*

He also expresses the opinion that the same may be true of the color photograph in Bechtel, Cribb & Launert (1981, 169).

*Ascocentrum curvifolium* has been by far the most frequently used *Ascocentrum* in the production of *Ascocenda* hybrids (*Ascocentrum* × *Vanda*). If *Asctm. rubrum* not only deserves status as a separate species but also has been used as a parent under the name *Asctm. curvifolium*, then our *Ascocenda* hybridizing records are hopelessly muddled. Was the *Ascocentrum* parent of the celebrated *Ascocenda* Yip Sum Wah (*Vanda* Pukele × *Asctm. curvifolium*), for example, really *Asctm. curvifolium*, or was it *Asctm. rubrum*? No recorded hybrids are attributed to *Asctm. rubrum*.

Only three of the *Ascocentrum* species have received any significant horticultural attention. These are *Asctm. curvifolium*, *Asctm. miniatum*, and *Asctm. ampullaceum*, listed in order of attention received. Christenson (1986b, 106–107) regards two other *Ascocentrum* species as having horticultural interest, but they are rarely seen in collections. These two are *Asctm. aurantiacum* (which apparently has been grown in some collections mislabeled as *Asctm. miniatum*) and *Asctm. pumilum*.

A brief horticultural description of individual *Ascocentrum* species follows, beginning with the three most commonly known in horticulture, listed alphabetically, and then proceeding to some of the others.

## Descriptions of *Ascocentrum* Species

### *Ascocentrum ampullaceum* (Roxb.) Schltr.

GENERAL LOCALE: Himalayan hills of India and eastward through Burma, the northwesternmost mountains of Thailand, Laos, and parts of southern Yunnan Province in China.

FLOWERING TIME: Spring. Requires coolness to initiate flowering.

DESCRIPTION: A compact plant, quite bushy as a result of its habit of producing many vegetative shoots from its base. Leaves are usually 13–15 cm (5–6 in) long and are closely spaced on the plant stalk. The foliage often is spotted, but this is not always the case; the amount of spotting is influenced by cultural conditions, especially the intensity of exposure to light. (See Plate 10.)

The erect inflorescences approximate the length of the leaves. The flowers usually are not held entirely above the foliage. The flowers measure about 2 cm (0.75 in) across. The inflorescence has up to 45 flowers, tightly but evenly arranged around and along the entire length of the flower stalk. They usually are rose-magenta in color; prized cultivars are a deep magenta color. The darkest form sometimes is called *moulmeinense* form. An orange-red form, *aurantiacum*, is also in cultivation, as is a rare pastel pink form. The species is variable; Seidenfaden

(1988, 316) remarks that perhaps it should be separated into several varieties or species.

## *Ascocentrum curvifolium* (Lindl.) Schltr.

GENERAL LOCALE: Assam, Burma, Thailand.
FLOWERING TIME: May, June. Requires coolness to initiate flowering.
DESCRIPTION: The plant grows taller than *Ascocentrum ampullaceum* or *Asctm. miniatum;* the plant stem can easily measure 50 cm (20 in) on a well-developed plant. The leaves are 15–25 cm (6–10 in) in length. The inflorescences also are 15–25 cm (6–10 in) long, erect, often held at least partly above the foliage. The flowers are well distributed along and around the stalk and measure nearly 2.5 cm (1 in) across, occasionally even more. On outstanding plants they may number up to 40, although the usual number is considerably less. The flowers are round and overlapping. The color is intense tangerine-orange or bright red; a concolor yellow form exists but is rare. (See Plate 11.)

Ascocentrum rubrum has been considered conspecific with *Asctm. curvifolium* by some botanists (see the discussion of the two species earlier in this chapter).

## *Ascocentrum miniatum* (Lindl.) Schltr.

GENERAL LOCALE: Thailand, Laos, Vietnam, northern Malaya and Java, at moderate altitudes of 240–750 m (800–2500 ft).
FLOWERING TIME: Early to mid-spring, but in cultivation may bloom at other times as well.
DESCRIPTION: A short, compact plant that flowers when quite young. A mature plant may grow to 30 cm (12 in) in height and has several basal or lateral shoots. The flowers are about 1.6 cm (0.6 in) across. They are densely clustered around the entire length of the inflorescence, which is erect and anywhere from 10–30 cm (4–12 in) long. Inflorescences with more than 50 flowers are common on awarded plants. Flower color ranges from very light yellow to orange. (See Plate 12.)

Christenson (1992) maintains that all Philippine records of this species are referable to *Ascocentrum aurantiacum,* and that the true *Asctm. miniatum* is an infrequently cultivated species from Java. The commonly cultivated plants from Indochina, which have been incorrectly called *Asctm. miniatum,* are a new species, *Asctm. garayi* E. A. Christ. Because of the uniform use of the name *Ascocentrum miniatum* in horticulture, however, and because the name probably will be conserved by the Royal Horticultural Society as a parent in the registration of hybrids, it is retained in this book. This is yet another example of the frequent discord between current botanical taxonomy and the practices followed in the registration of orchid hybrids. No feasible solution is in sight.

## *Ascocentrum aurantiacum* Schltr.
## *Ascocentrum aureum* J. J. Smith

*Ascocentrum aurantiacum* originally was described from Sulawesi (Celebes). The *Asctm. aurantiacum* plants available in commercial horticulture come from the Philippines, where the plants have been mistaken for *Asctm. miniatum*. Christenson (1986b) observes that *Asctm. aurantiacum* even was featured as *Asctm. miniatum* on the cover of the American Orchid Society *Bulletin* of April 1957. He describes *Asctm. aurantiacum* as intermediate between *Asctm. curvifolium* and *Asctm. miniatum*; it has the long, strap-leaf foliage of *Asctm. curvifolium* and the small, bright orange flowers of *Asctm. miniatum*. The vegetative stems of *Asctm. aurantiacum* are very short. The flowers are slightly smaller than those of *Asctm. miniatum* and are "pleasingly two-toned, highlighted by darker orange at the throat of the spur," according to Christenson. This desirable characteristic is worthy of introduction into *Ascocenda* breeding, he continues, but he then notes that the 20–25 cm (8–10 in) *Asctm. curvifolium*-type leaves and the short inflorescences that barely pass the top of the plant may be considered as liabilities. One recorded hybrid was made with *Asctm. aurantiacum*: *Ascocenda* Bill Fox, registered in 1975. The other parent was *Vanda* Rothschildiana.

*Ascocentrum aureum*, an obscure species, appears to be closely related to *Asctm. aurantiacum*; it differs principally by its smaller flowers. Seidenfaden (1988, 309), like Christenson, provisionally regards *Asctm. aurantiacum* and *Asctm. aureum* as two distinct species.

## *Ascocentrum himalaicum* Deb. Sengupta & Malick
## *Ascocentrum pumilum* (Hay.) Schltr.
## *Ascocentrum semiteretifolium* Seidenf.

Three possibly closely related ascocentrums bear semi-terete leaves: *Ascocentrum himalaicum*, *Asctm. pumilum*, and *Asctm. semiteretifolium*. Of these, only *Asctm. pumilum* is of horticultural interest.

*Ascocentrum himalaicum* has been found in northwestern India, Burma, and in part of the Yunnan Province of China. It is very closely related to *Asctm. semiteretifolium*. Little information is available, and it has attracted no horticultural interest.

*Ascocentrum pumilum* is a diminutive plant, less than 7.5 cm (3 in) tall, with similarly short semi-terete leaves. It bears 5–10 rose-pink flowers measuring about 1 cm (0.4 in) across, on a short inflorescence of about 2.5–3.8 cm (1–1.5 in). An awarded plant had 18 inflorescences when it was judged. The species does not have a definite flowering period and tends to bloom more than once each year. It would appear to be a delightful miniature. It is said to grow best under cooler and somewhat shadier conditions than the other ascocentrums. Christenson says it will thrive in a bright-light situation hanging above cool-growing genera such as *Masdevallia* and *Odontoglossum* that require less light. The species comes from Taiwan.

Two hybrids with *Ascocentrum pumilum* have been recorded, neither one of them involving a *Vanda*. One is *Angraecentrum* Rumrill Prodigy, registered in 1976, made with *Angraecum eichlerianum*. It is interesting to note that the genus *Angraecum* is in the Angraecinae subtribe of the Vandeae tribe, not in the Sarcanthinae subtribe, so this is a most unusual cross. The other hybrid with *Asctm. pumilum* is *Ascofinetia* Furuse, made with *Neofinetia falcata* and registered in 1979. The genus *Neofinetia* is in the Sarcanthinae subtribe.

*Ascocentrum semiteretifolium* was discovered in 1968 on an isolated mountaintop in northern Thailand (Seidenfaden 1988, 317) and later was refound in another part of the same region. It has semi-terete leaves, as does *Asctm. himalaicum*, and differs from the latter mainly in the shape of the flower spur. The spur of *Asctm. semiteretifolium* from its base curves forward under the lip, and is 5–7 mm (0.2–0.3 in), long, while the spur of *Asctm. himalaicum* is twice as long and at its base is parallel with the ovary; it then curves forward (Seidenfaden 1968, 317). The flowers are violet color, except for the side lobes of the lip, which are greenish yellow. The flowers have dark purple front edges.

### Hybrids within the Genus *Ascocentrum*

Only three primary hybrids within the genus *Ascocentrum* have been recorded. The first is *Asctm.* Sagarik Gold, made by Rapee Sagarik of Thailand and registered in 1966. It is a cross of *Asctm. curvifolium* and *Asctm. miniatum*. It has attractive, rather small, round flowers and has been widely distributed. The second primary hybrid is *Asctm.* Mona Church, made in Florida and registered in 1972. Its parents are *Asctm. miniatum* and *Asctm. ampullaceum*. The third is *Asctm.* Khem Thai, made in Thailand and registered in 1980. It is a cross of *Asctm. ampullaceum* and *Asctm. curvifolium*. It is variably colored in watermelon shades.

Thus, all of the possible primary combinations of these three widely distributed *Ascocentrum* species have been made. In addition, *Asctm.* Sagarik Gold (*Asctm. curvifolium × Asctm. miniatum*) was crossed back onto *Asctm. miniatum* to make *Asctm.* Sidhi Gold, registered in 1976. While additional *Ascocentrum* hybrids using some of the rarer *Ascocentrum* species are possible, it is safe to assert that the main contribution of *Ascocentrum* to orchid hybridizing will continue to be in combination with plants in other Sarcanthinae genera, and overwhelmingly with plants in the genus *Vanda*.

# ORIGIN AND IMPORTANCE OF THE GENUS *ASCOCENDA*

*Ascocenda* is a man-made genus; it is an intergeneric combination containing some ancestry from the genus *Ascocentrum*, some from the genus *Vanda* (it matters not how much), and none from any other genus. Both parent genera belong in the subtribe Sarcanthinae of the Vandeae tribe, as do *Aerides, Arachnis, Phalaenopsis, Renanthera, Rhynchostylis,* and *Vandopsis,* among others.

Vandas and ascocentrums, separately and together as ascocendas, have been crossed with many other members of the Sarcanthinae subtribe. While the genus *Vanda* has a greater number of intergeneric combinations to its credit than does any other natural genus—some 76 as of the end of 1992 (see Appendix D)—the 65 intergeneric combinations involving *Ascocentrum* place it in a tie with *Cattleya* for second place. Half of the intergeneric combinations with *Ascocentrum* also involve *Vanda* as a component, most often by way of an *Ascocenda* (*Ascocentrum* × *Vanda*) as a direct parent, rather than by a more circuitous route.

No other intergeneric combination with *Vanda* has anything like the widespread popularity of ascocendas. By year-end 1992, more than 1200 *Ascocenda* grexes had been registered. More ascocendas are grown in the United States than vandas. In the period 1984–92, the number of ascocendas receiving awards from the American Orchid Society considerably exceeded the number received by vandas.

## The Qualities of Ascocendas

Many of the ascocendas being exhibited today are hard to distinguish from vandas. Their flowers look like vandas in shape and often approach their size. When ascocendas first began to appear on the market in large numbers in the late 1960s, they often colloquially were called "miniature vandas" because of their smaller flower size. In most of today's ascocendas, however, the number of *Vanda* ancestors considerably exceeds the number of *Ascocentrum* ancestors, and, as a result, the descendants often approximate vandas in size and shape, and even in the smaller number of flowers. That is not always so, but it seems increasingly to be the case. Not everyone is of the opinion that this has been a desirable development. Some enthusiasts think that the smaller, more brightly colored, and more numerous flowers of the earlier *Ascocenda* hybrids, such as *Ascda.* Meda Arnold and *Ascda.* Yip Sum Wah, have a special charm lacking in some of the large-flowered ascocendas. Fortunately, most collections have room for both types.

The most outstanding quality of ascocendas indisputably is the combination of intense color and sparkling texture. An increased number of flowers is another desirable quality of ascocendas, even though this gain is accompanied by a reduction in flower size relative to that of vandas. The arrangement of the flowers is superior, on the whole. Roundness and flatness of flower shape are further qualities transmitted by ascocentrums. Their sepals and petals are of approximately the same size, and, while their flowers are slightly cupped overall, the individual parts are not twisted or recurved, unlike the flowers of many of the *Vanda* species. Plates 49 through 63 illustrate these fine qualities.

The pedicels of *Ascocentrum* blossoms tilt upward, displaying the individual blooms attractively. This highly desirable characteristic often is passed on to the *Ascocenda* hybrids and overcomes the tendency toward overly long pedicels and downward-tilting flowers that is evident in some vandas—an undesirable trait inherited from certain of their species-ancestors.

Vandas sometimes do not have as straight and erect a flower stalk as one would wish, whereas ascocentrums generally have straight, erect stalks and often transmit this fine feature to their *Ascocenda* hybrid progeny.

Finally, ascocendas bloom on rather small, compact plants, and at an early age. The cultural requirements of ascocendas are the same as those described for vandas in Part II of this book, and the plants need no special treatment. They are easy to cultivate and bloom in nearly all situations, provided they are given a reasonable amount of warmth, humidity, and light. Every greenhouse should include some ascocendas.

## THE *ASCOCENTRUM* SPECIES USED IN MAKING ASCOCENDAS

Only two of the *Ascocentrum* species have been used to any significant extent in making ascocendas: *Asctm. curvifolium* and *Asctm. miniatum. Ascocentrum curvifolium* has been used far more often than *Asctm. miniatum;* it has larger, rounder flowers and it breeds easily with vandas. *Ascocentrum miniatum* is a difficult breeder, and it is dominant in transmission of color—nearly all of its immediate progeny will bear orange or orange-yellow flowers. That can be an advantage or disadvantage, depending on the goals of the hybridizer.

The first *Ascocenda* on record is *Ascda.* Portia Doolittle (*Ascocentrum curvifolium* × *Vanda lamellata*). It was registered in 1949 by C. P. Sideris of Honolulu. In the following year, 1950, Sideris registered *Ascda.* Meda Arnold, a cross of *Asctm. curvifolium* with *V.* Rothschildiana. *Ascocenda* Meda Arnold is second only to *Ascda.* Yip Sum Wah (*V.* Pukele × *Asctm. curvifolium*) in influence in the development of award-winning ascocendas.

In addition, in 1951, Sideris registered the first *Ascocenda* made with *Ascocentrum miniatum.* As in the case of *Ascda.* Portia Doolittle, the *Vanda* parent was *V. lamellata.* He named the cross *Ascda.* Chryse. Since then, a number of ascocendas have been created with *Asctm. miniatum,* but not nearly so many as have been made with *Asctm. curvifolium.*

Very few ascocendas have been made with *Ascocentrum ampullaceum*. In 1970, *Ascocenda* Baby Blue was registered, a cross of *Asctm. ampullaceum* with *Vanda coerulescens*, by Richard Vagner of Florida. *Ascda.* Baby Blue apparently never was used in subsequent breeding of ascocendas, so it has contributed nothing to the development of the modern *Ascocenda*; however, one cultivar did receive an Award of Merit in 1989, the first and only imprint this hybrid has made on the scroll of *Ascocenda* history, apart from its being the first *Ascocenda* with *Asctm. ampullaceum* as a parent.

Only two other ascocendas have been recorded with *Ascocentrum ampullaceum* as a parent: *Ascocenda* Wichot, made in 1976 (with *Vanda bensonii*), and *Ascda.* Pink Thing, in 1977 (with *Ascda.* Rose Seidel). A number of intergeneric hybrids, however, have combined *Asctm. ampullaceum* with other Sarcanthinae genera (both natural and man-made), including *Aerides, Ascofinetia, Neofinetia, Neostylis, Renanthera, Rhynchostylis, Sarcochilus,* and *Vandofinetia.*

There are two reasons why *Ascocentrum ampullaceum* has not been used more extensively in the making of ascocendas. First is that the plant is notoriously difficult to breed with vandas. One experienced hybridizer in Thailand, on repeated occasions over a five-year period, tried to do so without success. Although he obtained seedpods in some instances, they always withered before maturing.

The second reason is that the inflorescences of *Ascocentrum ampullaceum* tend to be rather short and not stand out above the foliage sufficiently, a trait that makes some hybridizers uninterested in even trying to use this species as a parent. Nevertheless, more sustained efforts should be made with some of the outstanding line-bred cultivars of *Asctm. ampullaceum*. They have fairly large, well-shaped, deep magenta flowers that might produce some remarkable hybrids if successfully combined with good modern vandas.

One *Ascocenda* has been made with *Ascocentrum aurantiacum, Ascda.* Bill Fox (*Asctm. aurantiacum* × *Vanda* Rothschildiana). It was registered in 1975.

## THE LINEAGE OF THE AWARD WINNERS

The genealogy of recently awarded *Ascocenda* hybrids provides some basis for informed predictions about the directions most likely to lead to the award-winning grexes of the near future. The 265 *Ascocenda* plants receiving flower awards in the period 1985–92 (and published in the *Awards Quarterly* of the American Orchid Society no later than the fourth quarter of 1992) represented a total of 132 grexes. Of this total, 114 had some *Ascocentrum curvifolium* ancestry in their background and 30 had some *Asctm. miniatum* ancestry (13 had both). Only one (*Ascda.* Baby Blue) had any *Asctm. ampullaceum* ancestry.

Of the 132 grexes, only 24, or about 18%, had an *Ascocentrum* as an immediate parent; the lineage of the rest must be traced to their grandparents (usually) or beyond (less often) to find an *Ascocentrum* on their genealogical chart.

One *Ascocenda*, *Ascda*. Yip Sum Wah (*Vanda* Pukele × *Ascocentrum curvifolium*), created by Hawaiian hybridizer Roy Fukumura and registered in 1965, unquestionably has been the dominant progenitor of the best ascocendas. See Plate 49. Nearly 40% of the 265 awarded plants, and 26% of the grexes represented, had *Ascda*. Yip Sum Wah as a direct parent. Some others had *Ascda*. Yip Sum Wah as a grandparent. This degree of influence must be unique in the history of orchid hybridizing. Fukumura's name surely will be inscribed in any Orchid Hall of Fame.

Most of the *Ascocenda* Yip Sum Wah parents used in hybridizing were progeny of crosses of one *Ascda*. Yip Sum Wah clone by another *Ascda*. Yip Sum Wah clone, rather than remakes of *Vanda* Pukele × *Ascocentrum curvifolium*. Some of the offspring were polyploids.

The vandas used in producing the awarded ascocendas cover a wide range; they include *Vanda* species as well as hybrids. One hybrid, *Ascocenda* Amelita Ramos (*Ascda*. Pokai Victory × *V. sanderiana, alba*), received 10 flower awards during the 1985–92 period, of which 7 were Awards of Merit. Since one of its parents, *Ascda*. Pokai Victory, itself has *V. sanderiana* as a parent, the input of *V. sanderiana* was heavy (i.e., three out of the four grandparents of *Ascda*. Amelita Ramos). *Ascocenda* Amelita Ramos does not have *Ascda*. Yip Sum Wah in its pedigree, which proves that it is not absolutely necessary to have the latter as an ancestor. *Ascocenda* Amelita Ramos also demonstrates that *V. sanderiana* is as valuable in producing outstanding ascocendas as it is in producing outstanding vandas, which should come as no surprise.

*Vanda coerulea* has been used less often than *V. sanderiana* in the making of ascocendas, but its use has been very substantial, nonetheless. *Vanda coerulea* has made possible the creation of blue, purple, and pink ascocendas, and, in many cases, it produces interesting blue tessellation in the offspring. It also produces longer inflorescences than are found on ascocendas with no *V. coerulea* ancestry.

Without detracting from the influence of *Vanda sanderiana* and *V. coerulea,* it should be noted that some of the other *Vanda* species have been successful progenitors of award-winning ascocendas. In fact, the range is broader than in the case of awarded *Vanda* hybrids. *Vanda bensonii, V. coerulescens,* and *V. denisoniana* each served as a parent of two awarded *Ascocenda* grexes, *V. lamellata* of three, and *V. tessellata* of four. It may well be that some of the frustrated *Vanda* species will be more successful in producing future award-winning ascocendas than they have been in producing prize-winning *Vanda* hybrids, partly because the standards for ascocendas are slightly less stereotyped than the standards for vandas, which are dominated by the influence of *V. sanderiana*. See Plates 57 and 58 for a hybrid with *V. cristata* (which is shown in Plate 20).

The outstanding *Vanda* hybrids of the past decade, such as *V.* Kasem's Delight,

*V.* Charungraks, and *V.* Kultana Gold, also have been the most successful *Vanda* parents of ascocendas. The most frequently awarded *Vanda* hybrid in the first half of the 1980s was *V.* Kasem's Delight (Plate 33), and the most frequently awarded *Ascocenda* was *Ascda.* Yip Sum Wah (Plate 49). Together they bred *Ascda.* John De Biase (Plate 50), a hybrid that has garnered many flower awards of its own. It will be interesting to see how *Ascda.* John De Biase fares as a parent in the future. Crosses with this hybrid will be difficult to make because a tetraploid *Ascda.* Yip Sum Wah was the original (pollen) parent; most, and perhaps all, of the good off-spring are triploids, and therefore are likely to be sterile. Surely, many crosses will be attempted, however, and a few may be successful.

By the same line of reasoning, *Ascocenda* Fuchs Royal Dragon (*Vanda* Fuchs Delight × *Ascda.* Yip Sum Wah), registered in 1988, may produce some outstanding progeny. One cultivar of that grex, 'Indigo',  was awarded a First Class Certificate by the American Orchid Society in 1991. While both parents of this grex are exceptionally good in their own right and have proven to be very successful breeding stock individually, their offspring grex *Ascda.* Fuchs Royal Dragon may be another example of triploidy and sterility, because the *Ascda.* Yip Sum Wah parent used in making *Ascda.* Fuchs Royal Dragon was a tetraploid plant, just as was the case in the making of *Ascda.* John De Biase.

One should always bear in mind that the amount of recognition any orchid grex receives depends not just on the inherent quality of the members of the grex but also on the number of times the cross is repeated, and on the extent of the world-wide distribution of plants. Some outstanding hybrids of both vandas and asco-cendas have been made in Thailand and, for one reason or another, have never been exported to the United States or other countries where they could gain inter-national recognition. Some others have been exported in such small numbers from Thailand that the chances of a superb cultivar showing up in their midst is small, even if some remarkably good cultivars may exist in the country of origin.

## THE CHARACTERISTICS OF AWARDED ASCOCENDAS

Of all the considerable and desirable attributes of ascocendas, intensity of flower color and sparkling texture are the most valuable. An *Ascocenda* lacking in these two qualities has little advantage over today's best *Vanda* hybrids. By selective breeding over a number of generations, *Vanda* hybridizers did much to improve flower shape and arrangement and to produce straighter, more erect flower stalks and sturdy pedicels. But the introduction of *Ascocentrum* into the gene pools of their *Vanda* breeding stock supplied new potential for achievement that combinations of *Vanda* species alone could not accomplish.

In the awards descriptions of the American Orchid Society, the color and texture of awarded *Ascocenda* flowers are described with striking frequency as "sparkling," "crystalline," or "glistening." Along with saturated colors and a reflective texture, ascocentrums contribute bright orange, yellow, and red coloration to their *Ascocenda* hybrids, thereby attractively extending the range of colors and hues obtainable in *Vanda* hybrids. They often also contribute more pronounced dotting and speckling on the flowers. In short, they have contributed further excitement in several respects, which judges have appreciated.

The number of flowers per inflorescence of awarded ascocendas is spread over a wider range than in the case of vandas. Ascocendas displaying more of their *Ascocentrum* inheritance have more flowers than those that favor their *Vanda* ancestors. For *Ascocenda* flower awards published in the *Awards Quarterly* of the American Orchid Society from the beginning of 1989 through the end of the first quarter of 1993, the median number of flowers and buds on an inflorescence was 21. This compares with about 13 flowers per inflorescence for awarded vandas. In judging ascocendas, scoring on floriferousness has to be much more concerned with the plant's ancestry than is required in judging a standard *V. sanderiana–V. coerulea* type of *Vanda*. Much more variability exists among ascocendas, given the wide range of the proportion of *Vanda* ancestry.

This same dispersion is seen in flower size. If an *Ascocenda* has many vandas in its recent ancestry and, say, only one rather remote *Ascocentrum* ancestor, the flowers will be larger (and fewer in number) than if it had an *Ascocentrum* parent, or three *Ascocentrum* grandparents. In judging ascocendas, bigger is not always better. It is the combination of desirable traits, how each relates to the others, that gives an inflorescence dramatic presence—the "Wow! factor," to use a phrase coined by veteran orchid judge, hybridizer, and fancier, Benjamin C. Berliner. Or, as Supreme Court Justice Potter Stewart said about pornography, "I know it when I see it."

*Ascocenda* flower awards published in the *Awards Quarterly* of the American Orchid Society from the beginning of 1989 through the end of the first quarter of 1993 disclose that the median size of the flowers was merely 6.2 cm (2.4 in), but the spread around that figure was wide. A good general principle in judging ascocendas is that the fewer the number of flowers, the larger they should be, and vice versa. Even if the size of the flowers approaches that of awarded vandas, however, superiority is still expected in several other respects: the plant should have more flowers than a typical *Vanda*; the flowers should be better arranged on the flower stalk; the inflorescence should be erect; and, above all, the colors of the flowers should be bright and the texture crystalline. Such fine qualities should not be at the expense of frequency of blooming.

# 6

# Other Intergeneric Combinations with *Vanda*

## INTRODUCTION

Any grex with ancestry containing species of more than a single natural genus is placed in an intergeneric genus that combines the genera involved. An example is *Christieara* (*Aerides* × *Ascocentrum* × *Vanda*). To be placed in the genus *Christieara*, a hybrid must have among its ancestors at least one species of each of the three genera *Aerides*, *Ascocentrum*, and *Vanda*. The hybrid may have more than one ancestor from each genus. The remoteness and frequency of appearance of each ancestral genus and of the individual species within it do not matter. The result is an enormous number of possible combinations of species and of resulting hybrid traits, which is an enticement for hybridizers but an insurmountable obstacle for anyone trying to describe the features of the man-made genus in any detail—there is too much variability.

Another problem arises if the species composition of any genus is not defined uniformly by taxonomic authorities and by *Sander's List of Orchid Hybrids*. The problem is acute with respect to the genus *Vanda*, where a number of taxonomic changes have been made by botanists but have not yet been adopted for registration purposes by the International Registration Authority for Orchid Hybrids. In earlier years, *Vanda* species included two broad groups—those with strap-shaped leaves and those with terete leaves. (Hybrids combining the two types were called semi-terete vandas.) If an intergeneric combination contained a *Vanda* component, it mattered not whether the *Vanda* belonged in the terete or strap-leaf group; both were included in the *Vanda* genus as it was then defined by most botanists. Today, however, there is a divergence between the treatment by most taxonomists and the treatment in *Sander's List of Orchid Hybrids*—*Sander's List* still lumps them together, while botanists place the terete plants in other genera.

Two terete species cause especially bothersome problems because they have been used very extensively in hybridization with both strap-leaf vandas and with other members of the Sarcanthinae subtribe. These two are what now are widely identified as *Papilionanthe hookeriana* and *P. teres*, formerly described as *Vanda hookeriana* and *V. teres*. For registration purposes, *Sander's List* continues to treat the two species as vandas, although it recognizes that the correct botanical name is *Papilionanthe*. For the International Registration Authority to correct the situation, a large number of new intergeneric genera would have to be created—a daunting task.

In addition to *Ascocenda*, a combination covered in Chapter 5, 75 different intergeneric combinations with *Vanda* had been registered as of year-end 1992. These 75 intergeneric names are listed in Appendix D. The appendix also shows the number of hybrids that had been registered in each of the 75 genera, *after excluding those with terete-leaf ancestry*. The latter were excluded in order to make the definition of *Vanda* consistent with usage in the rest of this book—that usage is based on Christenson's list of the genus *Vanda* (see Appendix B), and it is in accord with current botanical taxonomy. The hodgepodge state of horticultural nomenclature and classification persuaded me to follow botanical rather than horticultural usage wherever a choice had to be made, unless the text indicates the contrary. Wherever *Sander's List* conflicts with that of Christenson and I have used Sander's nomenclature, I have placed the genus name in quotation marks. A taxonomically correct name was not available in such instances, because taxonomists do not give names to man-made hybrids of different species and genera—the only source is the International Registration Authority, on which *Sander's List* is based.

With one exception, all of the genera involved in the recorded combinations with *Vanda* belong in the Sarcanthinae subtribe of the Vandeae tribe. The sole exception is the genus *Aeranthes*, which is in the Angraecinae subtribe of Vandeae. One hybrid has been registered, *Vandaeranthe* Helmut Paul. It is a combination of *V. tessellata* and *Aerth. grandiflora*, originated by Helmut Paul of Hawaii. Paul has registered more than one intergeneric combination that no one else seems to have been able (or interested) to make.

All the species in the Sarcanthinae subtribe have 38 chromosomes in their somatic cells, with a few very rare exceptions (e.g., *Vanda* [=*Taprobanea*] *spathulata*). The cultivation of the species and hybrids described in this chapter is similar to that of vandas unless noted otherwise.

The number of man-made genera formed by combinations with *Vanda* is so large that it is necessary to be selective in describing the different ones and their respective constituent species. The criteria for selection are: (1) the number of grexes that had been registered in each of the man-made genera by year-end 1992; (2) the availability of plants from orchid nurseries and hybridizers; (3) the likely growth of popularity of a specific combination in the near future; and (4) in a few instances, the sheer novelty of a combination as an example of what can be accomplished by imaginative hybridizers. Clearly, these criteria involve a considerable amount of subjective judgment.

Some of the intergeneric hybrids are strikingly beautiful. Some have extended the limits of what had earlier been known and appreciated. Some have fragrance, a quality lacking in standard *Vanda* and *Ascocenda* hybrids. Some are very important commercially in certain parts of Southeast Asia. And finally, some offer exciting potential for hobby growers who want to venture beyond the present mainstream of *Vanda* and *Ascocenda* hybrids. For interior decoration and display purposes—a rapidly expanding use of orchids—some of the intergeneric combinations with *Vanda* are without peer.

In evaluating these types, the rather rigid value system used in the formal judging of standard vandas and ascocendas should be applied flexibly. In most instances, less emphasis should be placed on the size and shape of the individual flowers, and more weight should be given to the overall display and impact of the inflorescence—the number of flowers and their presentation, the brilliance of color, the texture, and possibly fragrance. The total effect should be evaluated not only up close but also at a distance. Distinctiveness also is an important consideration. For a steady diet, the conventional *Vanda sanderiana*–*V. coerulea* types that dominate the mainstream of *Vanda* and *Ascocenda* hybridizing may be preferred, but for the jaded appetite and the adventuresome, the intergeneric combinations offer an opportunity for sampling some uncommon delights.

Why are the intergeneric combinations with *Vanda* (other than *Ascocenda*) not better known and more widely available to hobby growers in the Western Hemisphere and Europe? Some of the plants, such as *Aranda* (*Arachnis* × *Vanda*), are too large for easy accommodation in most greenhouses. Some, such as *Arachnis*, require more intense light and more uniformly high temperatures than are practical in temperate climates, especially if the grower wants to have a diverse collection of orchids. These types are ideal for floral crops in tropical areas where they can be grown outdoors, and they have provided the foundation for a vast export trade in cut flowers.

Many of the intergeneric combinations are frustratingly difficult to make because of poor pairing of the parental chromosomes during meiosis, despite the fact that the chromosome count is almost completely uniform within the Sarcanthinae subtribe. When pods with viable seed are obtained, germination often is scanty and only a very small number of plants is produced—too few to make profitable cultivation and marketing feasible for larger nurseries. The owner of an internationally known orchid nursery in Thailand told me that he prefers to let amateur hybridizers "fool with them." Whenever a really good plant is achieved, he tries to buy one of the best of the cross and make mericlones by tissue culture. In that way, he can produce enough high-quality plants to warrant listing the grex in his catalog. Plants in the genus *Vascostylis* (*Ascocentrum* × *Rhynchostylis* × *Vanda*) were mentioned as one example.

Even when attempts at intergeneric combinations succeed in producing some offspring, the progeny frequently are sterile. As a result, hybridizers regard such crosses as a dead end as far as subsequent development is concerned. While the genera in the Sarcanthinae subtribe are related to one another, the disparities in the genes in many cases are so great that pairing of the chromosomes is complicated and diminished. The lack of promise discourages hybridizers from working toward improvement over an extended passage of generations, unlike the situation with vandas and ascocentrums, which breed together easily in most instances.

The main lines of intergeneric combinations with *Vanda* and *Ascocenda* (*Ascocentrum* × *Vanda*) are reviewed in this chapter not only because they represent an important contribution to floriculture, but also because some of the hybrids offer opportunities to expand orchid collections in enticing new directions.

### The Cut-Flower Trade—The Original Impetus

The cut-flower trade was the initial driving force behind the development of many of the intergeneric combinations with *Vanda*. After World War II, a large-scale, well-organized industry for producing and exporting cut orchid flowers developed in Southeast Asia, mainly in Singapore and Thailand, where favorable climatic conditions combined with substantial encouragement and support from official institutions.

At the beginning of the developmental period, some of the major commercial orchid growers in Hawaii were interested in moving in the same direction, but, partly because of inadequate government sponsorship and lack of cooperative organization (Kirch 1967, 193), their efforts did not thrive. The development of the orchid industry in Hawaii came to concentrate to a much greater extent on the local market for cut flowers and on the export market for potted plants; the latter do not require organized marketing on a large scale or as much uniformity of product.

The requirements of the Southeast Asian cut-flower industry were twofold. First, there was a need for frequent, economical air freight linking the growers' brokers with the major potential export markets—Japan, Western Europe, and North America. These facilities developed rapidly in the 1960s and 1970s. The second, parallel requirement was floriferous orchids that were easy to propagate and grow outdoors. The plants had to produce many-flowered inflorescences throughout the year, and the blooms had to be exotic and colorful enough for display purposes. In addition, the inflorescences had to pack and ship well and be long-lasting once unpacked. This very tall order was not fully met by the orchids available at the end of World War II. Hybridizers had their work cut out for them! They set to work. In Malaya, they were aided by encouragement and technical support from the Singapore Botanic Gardens.

From the outset, it was recognized that, along with some kinds of dendrobiums and, to a more limited extent, oncidiums, intergeneric combinations with *Vanda* (including *Papilionanthe hookeriana* and *P. teres*) offered much promise. Stem cuttings and tissue culture could provide ease of propagation and a uniform product.

The earliest center of the cut-flower industry was Singapore, although the future of that state as a producer currently is being threatened by the scarcity of land; in terms of volume, Thailand has superseded Singapore as the leading exporter. Exports of *Phalaenopsis*-type dendrobiums, such as *Dendrobium* Pompadour, outrank vandaceous orchids in Thailand, but Singapore remains an innovator and leader in intergeneric hybridizing, again thanks to the cooperation and influence of the Singapore Botanic Gardens, which also is making an effort to produce superior forms of some of the *Vanda* species (in contrast with *Vanda* hybrids, where Thailand leads). The existence of a profitable cut-flower industry provides a rationale for these efforts, even though the beneficiaries, in many instances, are, and probably increasingly will be, hobby growers worldwide.

## COMBINATIONS WITH *PAPILIONANTHE* AND *HOLCOGLOSSUM*

This book excludes from *Vanda* the terete-leaf orchids that botanists formerly placed in that genus. Nearly all taxonomists now regard what has been known as *Vanda teres* and *V. hookeriana* as more properly belonging in the genus *Papilionanthe*, and they place what has been called *V. amesiana* and *V. kimballiana*, along with several other species, in the genus *Holcoglossum*. See Christenson's list of the genus *Vanda* (in Appendix B) with respect to the foregoing. For hybrid registration purposes, however, *Sander's List* retains these four species in *Vanda*, and the International Registration Authority has not created new intergeneric names to cover combinations of either *Papilionanthe* or *Holcoglossum* with *Vanda*. Whenever a generic name is in conflict with current taxonomy, as reflected by Christenson's list, but reflects only the usage in *Sander's List*, I have placed the generic name in quotation marks, to alert the reader to the discrepancy.

As of year-end 1992, the hybrid *Papilionanthe* Miss Joaquim (*P. hookeriana* × *P. teres*) was listed (as a *Vanda*) as the direct parent of more than 50 hybrid "*Vanda*" grexes. "*Vanda*" Josephine van Brero (*V. insignis* × *P. teres*) had parented some 75 hybrid grexes; it had been combined with 10 strap-leaf *Vanda* species as well as with a much larger number of strap-leaf *Vanda* hybrids.

Intergeneric hybrids made with *Holcoglossum* have not been of much consequence for the cut-flower trade, nor are they widely grown by hobbyists, even in
the tropics, but that is not the case with *Papilionanthe* intergeneric hybrids. Such
hybrids, along with the intrageneric cross *P*. Miss Joaquim, have played a significant role in the cut-flower industry. These plants, like *Arachnis,* can be grown outdoors in full or nearly full sun, in rows on the ground (rather than in pots on
benches). Some of the semi-terete hybrids are famous. The most influential hybrid
may be *"Vanda"* Josephine van Brero (*V. insignis* × *P. teres*), which, by year-end
1992, had parented 112 crosses in the Sarcanthinae subtribe. Another famous hybrid is *"V."* Emma van Deventer (*P. teres* × *V. tricolor*). This latter hybrid, when
crossed with *V. sanderiana,* produced *"V."* Nellie Morley (Plate 64), a grex registered in 1952 by Harold Morley of Hawaii. The pod parent, *"V."* Emma van Deventer, was a tetraploid. So remarkable is the hybrid that it often is called "The
Pride of Hawaii."

*Papilionanthe teres* is a climbing, densely growing, scrambling plant that is
widely distributed in Thailand, Laos, Burma, and the foothills of the Himalayas.
It usually flowers from April to July. The inflorescence is only about 20 cm (8 in)
long and bears 2–5 flowers measuring 6.5–7.5 cm (2.5–3 in) across. The petals are
completely twisted around. The flowers usually have mauve petals, with darker
side lobes and midlobes. A very attractive white color form exists. The plants require strong light and high temperatures. They do not begin blooming until the
plants are large, which is detrimental to cultivation in greenhouses.

*Papilionanthe hookeriana,* the other *Papilionanthe* species that has played a
great role in hybridizing, is found in Malaysia, in the adjoining southern peninsular region of Thailand, and in Sumatra and Borneo. It has a scrambling habit and
is found in swampy areas. The plants do not bloom until they have reached considerable height. The inflorescence is rather like that of *P. teres,* but the flowers
generally are a bit smaller. The color of the flowers similarly is light mauve, but the
lip is much larger and is colored purple, with larger, darker purple blotches. The
flowers are far more striking than those of *P. teres* because of the larger, magnifi-
cently colored lip. As in the case of *P. teres,* a white flower form exists.

## SOME COMBINATIONS WITH *ARACHNIS*

*Arachnis* Blume ranges from Burma and Indochina down through the Malayan
Peninsula, Borneo, and into Indonesia and Papua, New Guinea. There are some
17 species, 3 of which are found on the Malayan Peninsula: *Arachnis hookeriana*
(Reichb. *f.*) Reichb. *f.*; *Arach. flos-aeris* (L.) Reichb. *f.*; and *Arach. maingayi* (Hk.
*f.*) Schltr. Perhaps because of their availability and suitability for cultivation locally, these three are the only ones that have been used for hybridization. *Arach-*

*nis maingayi* is considered to be a natural hybrid of the first two, but it generally is treated as a species in the literature. For a description and color photographs of a number of the species, see Lamb (1983, 175–176).

*Arachnis hookeriana* color form *luteola* is the form of *Arach. hookeriana* that has most often been used in breeding. *Arachnis hookeriana* produces progeny that flower at an early age and regularly—two highly desirable qualities. The form *luteola* has yellowish flowers with no purple dotting, unlike the standard form, which has white or cream-colored flowers with a pink or purplish lip and faint pink dotting. *Arachnis hookeriana* plants typically have long, upright, unbranched inflorescences bearing 4–8 widely spaced flowers, measuring about 6 cm (2.5 in) vertically and somewhat less horizontally.

*Arachnis flos-aeris* plants have even longer inflorescences than those of *Arach. hookeriana*—about 1.5 m (5 ft) long; they may be branched or unbranched, upright or drooping, with a highly variable number of flowers measuring about 10 cm (4 in) vertically and 7.5 cm (3 in) across. The flowers have pale yellowish green sepals and petals marked with prominent dark purplish brown, irregularly shaped bars and spots. The lip is a blend of brown and dull purple. As in the case of *Arach. hookeriana,* the column is white or creamy white. The flowers have a musky scent. *Arach. flos-aeris* var. *insignis,* sometimes called the black scorpion orchid, is strikingly beautiful. Its sepals and petals are dark maroon and look almost black at their tips. Its column is white, and the red lip has a patch of white on the midlobe.

Native habitats of the species are Sumatra, Java, Borneo, Malaya, and the Philippines.

*Arachnis maingayi* plants have inflorescences like those of *Arach. flos-aeris* and have flowers that are similarly shaped but slightly smaller. Flower color is whitish with a pink overlay, and the sepals and petals are rather densely barred and blotched with pink or light purple. The column is white.

The sepals and petals of the flowers of *Arachnis flos-aeris* and *Arach. maingayi* are narrow; their clubbed tips are prominently rolled forward, and the edges of the sepals and petals are curved backward. The flowers have considerable substance. The plants commonly are called the scorpion orchid, because their bowlegged lateral sepals are imagined to resemble a scorpion's claws, and the tall, narrow dorsal sepal its upright tail.

*Arachnis* plants do best in a uniformly hot climate. They are grown in full sun in Southeast Asia, and they grow rapidly. Under hot, sunny conditions, *Arachnis* plants produce a new leaf every two weeks, and up to 12 inflorescences per plant stem annually. The plants are very large and usually are grown in hedgerows or in clay pots placed on the ground and mulched.

At the outset, *Arachnis* formed the foundation of cut-flower production in Singapore. Its potential for intergeneric hybridizing intrigued two pioneer enthusiasts in Singapore, John Laycock, a founder of the Malayan Orchid Society in 1928 (now known as the Orchid Society of South East Asia), and R. E. Holttum, director of the Singapore Botanic Gardens from 1925 to 1949. They began crossing *Arachnis* with *Vanda* to produce *Aranda*. Thanks to their initiatives, hybrids combining the two genera became a mainstay of the cut-flower industry. The hybrids outdo their parents both in floriferousness and in intensity of color.

Extended hybridizing with the two genera was not easy, partly because their hybrid progeny are notoriously difficult to breed with one another; they tend to be sterile or to exhibit very low rates of germination. Notwithstanding the difficulties, the registered combinations of *Arachnis* with *Vanda* (including *Papilionanthe*) far outnumber all the other intergeneric combinations with the genus *Vanda*, except for *Ascocenda* hybrids, which lead by a wide margin. Some of the outstanding *Aranda* hybrids are discussed in the next section.

### *Aranda* (*Arachnis* × *Vanda*)

The combination of *Vanda* with *Arachnis* produces shorter inflorescences and more closely spaced flowers than are found on *Arachnis* species—in both respects an improvement over *Arachnis*—and a wider range of colors.

The early *Aranda* hybrids were developed mainly in Singapore and Malaysia, but one-fifth of the *Aranda* crosses registered in the period 1946–60 were originated by hybridizers residing in Hawaii. They, too, demonstrated interest in the potential of this intergeneric combination.

Two *Arachnis* species and two *Arachnis* intrageneric hybrids completely dominated the production of *Aranda* hybrids during the 1946–60 period. The two species were *Arach. hookeriana*, mainly the color form *luteola,* and *Arach. flos-aeris,* each of which parented about one-fourth of the total *Aranda* hybrids recorded. The two dominant *Arachnis* intrageneric hybrids were *Arach.* Ishbel (*Arach. maingayi* × *Arach. hookeriana*) and *Arach.* Maggie Oei (*Arach. hookeriana* f. *luteola* × *Arach. flos-aeris*), both registered in 1940.

In Singapore, these two intrageneric hybrids flower almost continuously throughout the year. *Arachnis* Ishbel was made by H. G. Grieve of Tebrau, Johore, Malaysia, but the grex was raised and registered by the Singapore Botanic Gardens. *Arachnis* Ishbel is free-blooming and fertile, and it has produced some excellent offspring, a few of which also have been fertile and therefore usable for subsequent breeding. *Arachnis* Maggie Oei (Plate 65) was originated by John Laycock. It was a major breakthrough because at least one cultivar, 'Red Ribbon', was tetraploid and parented some outstanding progeny. Just as tetraploids played a major role in the advancement of hybrids in many other genera of orchids, they also contributed much to the development and extensive use of *Arachnis* flowers in the cut-flower trade of Southeast Asia.

In addition to being of exceptionally high quality, *Arachnis* Ishbel and *Arach.* Maggie Oei were two of the earliest hybrids made within the genus *Arachnis*. In both cases, the seed was sown in 1935, but *Arach.* Ishbel first bloomed in November 1938, while seedlings of tetraploid *Arach.* Maggie Oei did not flower until November 1940 (Henderson and Addison 1956)—tetraploids, when young plants, grow more slowly than diploids. Both of these artificially produced hybrids parented numerous progeny; the cut-flower trade owes a great deal to them. On occasion, their tetraploid forms are used as parents even today.

The flowers of nearly all of the first-generation *Aranda* hybrids resemble those of *Arachnis* in shape, although the sepals and petals are a bit wider and flatter (see Plates 66–68). A notable exception is *Aranda* Gold Star (*Arach. hookeriana* × *"Vanda" spathulata*), which clearly shows the influence of the latter parent—another example of the overwhelming dominance of this hexaploid species, almost regardless of whatever it is crossed with. (Christenson places *"V." spathulata* in the genus *Taprobanea*.)

Two things about *Aranda* hybrids registered in the 1946–60 period stand out. The first is the influence of R. E. Holttum and the Singapore Botanic Gardens. As early as 1928, Holttum started propagating native orchids of Malaya in flasks, and he produced a number of hybrids, not only in the genus *Arachnis* but also in the genera *Phalaenopsis, Renanthera, Spathoglottis,* and *Vanda*. His guiding spirit continued to be felt even after his retirement as director of the botanic gardens in 1949, when he became professor of botany at the newly established University of Malaya in Singapore, a post he held until his return to England in 1954. He is known in Malaysia as the father of the Malayan orchid, a title he richly deserved. Of the *Aranda* hybrids attributed to Singapore origin, the vast majority were registered by the Botanic Gardens. It was Holttum's dream, it is said, to develop floriculture based on Malayan orchids. The record shows that he succeeded.

The second thing that stands out in the breeding of *Aranda* hybrids is the reliance on *Vanda* species, as contrasted with *Vanda* hybrids. Of the 48 *Aranda* crosses registered in the 1946–60 period, 28 had a species of strap-leaf *Vanda* as a parent, and 13 different *Vanda* species were employed. The most frequently used species was *V. tricolor* (including its variety *suavis*). It was a parent of 6 of the 28 grexes that had a *Vanda* species as a direct parent. The 48 *Aranda* crosses do not include two made with *Papilionanthe*. As already noted, *Sander's List* does not make the distinction between *Vanda* and *Papilionanthe* when a grex is registered.

New *Aranda* hybrids continued to be made in considerable number. In the 1981–92 period, 46 were registered. Most were made by breeding an *Aranda* (rather than an *Arachnis*) to a *Vanda*, thereby reinforcing the *Vanda* genes. The other hybrids, with two exceptions, did have an *Arachnis* parent; in the two exceptions, both parents were *Aranda*. This record probably is a reflection of the problem of finding *Aranda* plants that are fertile. The *Vanda* hybrids most frequently chosen for breeding had blue or purple flowers, it seems, from a review of the record of the *Vanda* parents used.

One *Aranda* has been a conspicuous parent, and it is the only one that has parented many grexes. It is *Aranda* Christine (*Arachnis hookeriana* × *Vanda* Hilo Blue), registered in 1963 by Federal Orchids of Malaysia. (See Plate 66.) In the 1981–92 period, it was combined with 18 different *Vanda* hybrids, as well as with many *Ascocenda* plants (to create *Mokara*). *Aranda* Christine plants, at least the cultivars used as stud plants, are tetraploids.

## *Mokara* (*Arachnis* × *Ascocentrum* × *Vanda*)

The main thrust of *Mokara* breeding has been to increase the floriferousness of *Aranda*, to widen the sepals and petals, to brighten the colors, and to achieve flowering at an earlier age. Flower size is reduced somewhat. Cut-flower production remains the major goal, and *Mokara* is increasingly important in the cut-flower trade.

The first hybrid registered in this trigeneric genus was *Mokara* Wai Liang, registered in 1969 by C. Y. Mok of Singapore. It is a cross of the famous *Arachnis* Ishbel with *Ascocenda* Red Gem (itself a stunning cross of *Vanda merrillii* and *Ascocentrum curvifolium*). In the following year, Mok registered *Mkra.* Ooi Leng Sun, another hybrid made with *Arach.* Ishbel, but this time with *Ascda.* Meda Arnold as the *Ascocenda* parent.

In the 1981–92 period, 47 new *Mokara* hybrids were registered, 21 of which had an *Aranda* as a parent; 9 of the crosses had an *Arachnis* parent. One *Aranda*, the earlier-mentioned *Aranda* Christine (Plate 66), accounted for 14 of those grexes having an *Aranda* parent, a further indication of the limited number of available *Aranda* that could be used successfully as stud plants. Seventeen of the 47 *Mokara* crosses registered during 1981–92 had one *Mokara* parent. Two *Mokara* hybrids were most popular as parents: *Mkra.* Khaw Phaik Suan and *Mkra.* Clark Kuan. None of the crosses used a *Mokara* for both parents. In every instance, *Ascocentrum* was introduced via an *Ascocenda* rather than directly. If the latter approach had been used, in all likelihood it would have reduced flower size more than was desired.

*Mokara* was developed primarily for the cut-flower trade, but it offers exciting potential for the hobby grower as well. By highly selective breeding of *Mokara* plants with *Ascocenda* plants, especially with repeated injection of *Ascocenda* parentage, it should be possible to reduce plant size further and to retain the best qualities of *Mokara*, such as a wide range of vibrant colors, long inflorescences bearing many flowers of extraordinary substance, and a free-blooming habit. The main restricting factors are limited fertility and the difficulty of producing adequate quantities of siblings; however, persistence may overcome this problem in enough instances to make efforts worthwhile. Response to this challenge probably will have to come from determined amateurs and dedicated commercial nurseries, such as Lum Chin Orchids in Kuala Lumpur, Malaysia. That nursery was responsible for the great majority of the 22 *Mokara* hybrids registered in the years 1981–85. For a photo of an excellent *Mokara*, see Plate 69.

## SOME COMBINATIONS WITH *AERIDES*

Species of *Aerides* Loureiro are scattered throughout tropical Asia. (For a succinct and current review of the taxonomy of this genus, see Christenson [1993].) The genus contains many showy and floriferous species. The flowers are well arranged on long, arching or pendulous stalks. The individual flowers have appealing features, such as fascinating lips with prominently protruding spurs and complex combinations of flower colors. The texture of the blooms is glistening and waxy, and their substance is heavy. Fragrance is another desirable quality. One unattractive and limiting trait, however, is that the *Aerides* species that have most frequently been used in intergeneric breeding tend to be large and rather coarse-appearing plants, and they are rambling in their habit of growth.

Like vandas, *Aerides* require bright, though diffuse, light if they are to bloom well. They are not difficult to grow under greenhouse conditions in temperate climates, but they must be provided with warm moisture and generous air movement. Without good movement of air, they often succumb to rot.

As can be seen in Appendix D, a number of intergeneric crosses have been made combining strap-leaf *Vanda* with *Aerides*—either with plants in these two genera alone or in combination with one or more additional vandaceous genera. The *Aerides* species that have been used to make intergeneric hybrids with *Vanda* and *Ascocenda* include: *Aer. lawrenceae* Reichb. *f.*; *Aer. crassifolia* Par. ex Burbidge; *Aer. falcata* Lindl., including the variety *houlletiana* (Reichb. *f.*) Veitch, which some taxonomists (and *Sander's List*) maintain as a distinct species; *Aer. multiflora* Roxb. (Plate 70); *Aer. quinquevulvera* Lindl.; *Aer. odorata* Loureiro; and *Aer. leeana* Reichb. *f.* To this list should be added *Aer. krabiensis* Seidenf., which seems to have considerable potential for hybridizing, but which, as of year-end 1992, had not been registered as a parent of a hybrid.

Two other species also have been registered as *Aerides* parents. They are *"Aer."* *flabellata* (Rolfe ex Downie) E. A. Christ. (see Christenson's list of the genus *Vanda* in Appendix B) and *Seidenfadenia mitrata* (Reichb. *f.*) Garay. Christenson regards the proper generic placement of *"Aer."* *flabellata* as unresolved, but in any case not belonging in the genus *Aerides*.

Two intergeneric combinations involving *Vanda* and *Aerides* have been made in some quantity: one is *Aeridovanda* (*Aerides* × *Vanda*); the other is *Christieara* (*Aerides* × *Ascocentrum* × *Vanda*). Of the two, *Christieara* hybrids have been registered in greater number, but *Aeridovanda* hybrids have been much more prominent in the cut-flower trade. Both combinations, however, depending on the composition of the *Aerides* ancestry and the related growth habit, can produce hybrids suitable for the hobby grower.

### *Aeridovanda* (*Aerides* × *Vanda*)

Through the end of 1992, 96 *Aeridovanda* crosses had been recorded in *Sander's List of Orchid Hybrids*. The *Aerides* species most frequently used as a parent was *Aer. lawrenceae*, a tall Philippine species that usually carries more than 20 flowers, well arranged on a hanging, unbranched inflorescence. In its native habitat on the island of Mindanao, the species blooms in the period August through November. The flowers are the largest in the genus and measure slightly more than 2.5 cm (1 in) across. They are pleasingly fragrant. Typically the flowers are waxy white, or creamy white, with a prominent amethyst-purple blotch on the tips of the sepals and petals; the lip has reddish purple highlights and a very conspicuous, protruding spur that adds much to the charm of the flowers. The grandness of the flowers undoubtedly explains why *Aer. lawrenceae* clones have been used as a parent so often in the making of *Aeridovanda* hybrids, despite the fact that its flowers are rather short-lived. Unfortunately, this fault tends to be passed on to its intergeneric progeny. Of the 96 *Aeridovanda* hybrids registered up to the end of 1992, 30 had *Aer. lawrenceae* as a direct parent. The hybrids come in a number of attractive colors (see Plates 71–73).

The hybridization records do not indicate any great interest on the part of the cut-flower industry of Southeast Asia in producing *Aeridovanda* crosses based on *Aerides lawrenceae*, perhaps because of the problem of longevity of the flowers, combined with seasonality of blooming. Registrants from Hawaii, where breeders are less influenced by considerations of suitability for the cut-flower export trade, accounted for slightly more than half of the 30 *Aer. lawrenceae*–*Vanda* crosses recorded through 1992. Eight hybrids were registered prior to 1961; 7 of them by breeders located in Hawaii, and 1 by E. S. Wright of San Antonio, Texas. None were registered by breeders located in the creative center of the Southeast Asian cut-flower industry, Singapore and Malaysia; however, hybridizers from that area did register 3 *Aer. lawrenceae*–*Vanda* hybrids in the 1961–70 decade, and another 4 in the subsequent 20 years—hardly a groundswell of enthusiasm for this line of breeding.

Six of the 96 *Aeridovanda* crosses used the hybrid *Aerdv.* Vieng Ping as a parent, an exception to the general practice of using only *Aerides* species as the vehicle for introducing *Aerides* genes. This is a tribute to the fertility of *Aerdv.* Vieng

Ping. *Aeridovanda* Vieng Ping is a Thai hybrid registered in 1973; it has *"Aer."* *flabellata* (whose placement in this genus is challenged by Christenson) as one of its parents and *Vanda denisoniana* as the other. Its flowers are yellow and rather closely resemble those of the *V. denisoniana* parent in appearance. *Vanda denisoniana* has been used as a direct parent of an *Aeridovanda* hybrid a total of 6 times, and another 6 times as a grandparent via *Aerdv.* Vieng Ping.

Apart from *Aeridovanda* Vieng Ping, only 4 *Aeridovanda* hybrids had served as a parent of another *Aeridovanda* by year-end 1992, for a total of 5 individual grexes. Inasmuch as 96 hybrids were registered in the 47-year period 1946–92, the limited use of *Aeridovanda* hybrids as stud plants suggests that fertility is a major problem with most members of this intergeneric combination, a conclusion that is supported by my discussion of the matter with experienced breeders.

All in all, *Sander's List of Orchid Hybrids* discloses that 11 different *Aerides* species have been used as a direct parent of an *Aeridovanda*. We already have mentioned *Aer. lawrenceae*, which was used 30 times, and *"Aer."* *flabellata*, although not properly belonging in this genus, was used on 7 occasions. *Aerides odorata*, a very tall and widely distributed species, is recorded as a parent in 16 instances (8 of which were attributed to Hawaiian breeders). *Aerides quinquevulvera* and *Aer. leeana* (as *Aer. jarckiana* Schltr.), both of which are broad, short plants, each was used 8 times. *Aerides crassifolia*, a compact Thai species with mauve flowers, was a parent 7 times. The remaining 5 species each was used from 1 to 4 times.

Thus, while it is demonstrable that there has been a broad experimentation with *Aerides* species, the number of occasions involving any one of them has been limited, save for *Aer. lawrenceae*. The possibilities for combining *Aerides* species with *Vanda* species and hybrids seem far from exhausted, since more than a few attempts, and with different mates, are needed before one can draw reliable conclusions about the potential of any particular line of intergeneric breeding. Nonetheless, for hobby growers, it is likely that the future lies more with *Christieara* than with *Aeridovanda*.

### *Christieara* (*Aerides* × *Ascocentrum* × *Vanda*)

*Christieara* is a combination of *Aerides*, *Ascocentrum*, and *Vanda*, in any order and frequency of representation of each. In producing *Christieara*, however, hybridizers almost always have used an *Ascocenda* parent with an *Aerides* parent, rather than crossing an *Aeridocentrum* (*Aerides* × *Ascocentrum*) with a *Vanda* parent, or combining an *Aeridovanda* with an *Ascocentrum*. Either of these other two routes would have been possible, but the outcomes would not be identical in terms of the characteristics of the offspring.

The principal reasons for adding *Ascocentrum* genes to those of *Aerides* and *Vanda* have been to attain a wider range of bright colors and a larger number of flowers. Possibly, earlier age of first-blooming and greater frequency of flowering were other considerations. One might think that reducing plant size would have been a goal, too, but the frequency of use of large-sized *Aerides* species as parents suggests that this consideration was not high on the list of priorities of most of the hybridizers. Plant size generally is not of great concern to breeders in the tropics, where most plants are grown outdoors or with minimal enclosure.

By year-end 1992, a total of 61 *Christieara* hybrids had been recorded. The first was registered by Welda F. Christie of Miami, in 1969. The hybrid was *Chtra.* Lillian Arnold, a cross of *Aerides lawrenceae* by *Ascocenda* Meda Arnold. In the decade 1971–80, that landmark cross was followed by 22 others, utilizing plants of 11 separate species of *Aerides* as parents. By the end of 1992, a further 38 *Christieara* hybrids had been registered. Two fine *Christieara* hybrids, *Chtra.* Renee Gerber and *Chtra.* Fuchs Confetti, are shown in Plates 74 and 75.

In the making of *Christieara* hybrids, a preference for the smaller-sized species of *Aerides* as parents might have been expected, but that was not the case. Forty-five of the total of 61 crosses used an *Aerides* species as one of the direct parents, and 18 of these used *Aer. lawrenceae,* a very tall, rambling species. Its share was nearly four times that of the next most numerous species parent, *"Aer." flabellata,* which was used in 5 cases. Since *"Aer." flabellata,* regardless of its proper generic placement, is quite a short plant and, in addition, has a particularly attractive wide, fringed lip, one would have expected it to be more popular than it was, despite its bearing fewer flowers and having a shorter raceme. Its flowers are not long lasting, but neither are those of *Aer. lawrenceae.* Following in frequency of usage as parents came *Aer. odorata, Aer. quinquevulvera,* and *Aer. falcata* var. *houlletiana,* all of which are long-stemmed plants when mature. Two smaller *Aerides* species, *Aer. multiflora* and *Aer. crassifolia,* were direct parents only 3 and 2 times, respectively.

It is a pity that more hybridizing has not been done with the shorter-stemmed *Aerides,* such as the foregoing two (*Aer. multiflora* and *Aer. crassifolia*), along with *"Aer." flabellata* and the more recently available *Aer. krabiensis.* They could help *Christieara* hybrids to reach more fully the potential of this man-made genus; namely, compact plants with many brightly colored flowers well arranged on either a branched or unbranched inflorescence. It should be noted, however, that both *Aer. multiflora* and *Aer. krabiensis* lack fragrance. And it also should be noted that *Aer. multiflora,* even the lowland variety *godefroyae,* do not flower well (if at all) in Singapore; but that should not have discouraged breeders in some other places.

A description of one awarded plant, *Christieara* Manoa 'Robert', AM/AOS, illustrates the great potential of *Christieara* hybrids. This hybrid is a cross of *Aerides multiflora,* a short-stemmed species of erect habit, and *Ascocenda* Yip Sum Wah:

Forty-five flowers and six buds on one inflorescence: flowers . . . deep rosy pink with a flush of apricot at base of petals, maturing to apricot with a flush of rose pink on sepals and apices of petals; color changes gradual and pleasing . . . texture sparkling; flowers closely arranged around inflorescences . . . Natural spread of flower 3.7cm (1.46 in).

Twenty-three *Christieara* cultivars, representing 14 grexes, had flower awards published by the American Orchid Society in the 1981–92 period, a large number considering the relatively sparse distribution of the genus. The record attests to their appealing qualities. More awarded cultivars will undoubtedly follow, as hobby growers add this delightful trigeneric to their collections and exhibit the plants at shows and judging sessions. The plants can be grown under the same conditions that are optimum for vandas and ascocendas.

## Lewisara (Aerides × Arachnis × Ascocentrum × Vanda)

*Lewisara* is a seldom encountered multigeneric comprised of *Aerides, Arachnis, Ascocentrum,* and *Vanda.* Only 12 grexes of this complex man-made genus had been recorded as of the end of 1992, but some of the cultivars are of exceptional merit. *Lewisara* is an exciting line of breeding with considerable promise for the hobby grower.

The first 2 crosses registered were *Lewisara* Gracia and *Lwsra.* Max, both by M. Yamada of Hawaii, in 1967. The genus is named after Gracia Lewis of Singapore, an active vandaceous hybridizer in her own right. The second cross was named after her husband, Max. The parentage of these *Lewisara* hybrids, in both instances, consisted of *Aeridachnis* Bogor (*Arachnis hookeriana* × *Aerides odorata*) on the maternal side and an *Ascocenda* on the other side. The *Arach. hookeriana* used in making *Aerdns.* Bogor was not the color form *luteola,* but a form with attractive, rather small, white flowers. From the hybridizer's viewpoint, it also had the great advantage of being a good breeder.

The next grex was not registered until 1975. It, too, had *Aeridachnis* Bogor as the pod parent and an *Ascocenda* (Blue Boy) plant as the pollen parent. *Ascocenda* Blue Boy resembles a miniature *Vanda coerulea,* and therefore is a highly desirable grex. The hybrid was named *Lewisara* Chittivan by its Thai originator. This *Lewisara* hybrid has round, flat, reddish purple tessellated flowers clearly reminiscent of the pollen parent. See Chitanondh (1987, 177) for a color photo of *Lwsra.* Chittivan. Three more *Lewisara* grexes were recorded near the end of the 1970s, 2 of which also had *Aerdns.* Bogor as their pod parent; the third had *Aerdns.* Bogor as a maternal grandparent. All of them had an *Ascocenda* as the pollen parent, and all 3 hybrids originated in Singapore. Finally, in the period 1981–92, 6 more *Lewisara* were registered, and every one of them likewise had *Aerdns.* Bogor either as its mother or as its maternal grandparent. This is quite an astonishing record; of the 12 *Lewisara* grexes registered to date, all are based on *Aerdns.* Bogor.

The other interesting genealogical note is that *Lewisara* Chittivan, the 1975 Thai hybrid, was the pod parent in 4 instances. It is an attractive hybrid by itself, and a good breeder as well, a trait inherited from its own pod parent, *Aeridachnis* Bogor. In 3 of these 4 instances, the pollen parent was a *Vanda*; in the remaining one, it was the celebrated tetraploid *Aranda* Christine.

An outstanding grex is *Lewisara* Fatima Alsagoff (*Lwsra.* Chittivan × *Vanda* Wirat), originated by Multico Orchids of Singapore in 1982. It has round, flat flowers that are a strikingly beautiful, uniform violet-purple color. They measure 7 cm (2.75 in) across and are borne on a strong, upright inflorescence with 15 or more well-arranged flowers. Flower substance is firm. The plant is easy to grow, prefers somewhat shaded areas, and is free-flowering. It is reported to have first bloomed almost exactly five years from the date of pollination, so the cross obviously can bloom at a relatively early age. A photograph of the cultivar 'Zahrah' is presented in Plate 76.

## SOME COMBINATIONS WITH *RHYNCHOSTYLIS*

*Rhynchostylis* Blume is a genus containing three species, each of which has considerable horticultural importance: *Rhy. coelestis* Reichb. f. (Reichb. f.); *Rhy. gigantea* (Lindl.) Ridl.; and *Rhy. retusa* (L.) Bl. All of them are widespread in Thailand. All have fragrant flowers. The plants are stout and quite short. *Rhynchostylis coelestis* has an erect inflorescence, while *Rhy. gigantea* and *Rhy. retusa* have long, pendent, unbranched racemes bearing an abundance of flowers tightly arranged cylindrically. This characteristic accounts for these latter two species sometimes being called the fox-tail orchids.

To thrive and flower well, *Rhynchostylis* plants must have a wide spread between daytime and nighttime temperatures; they will not tolerate the uniformly high temperatures of places like Singapore. Moreover, they grow best with less intense light than is optimum for vandas and ascocendas. The plants do well in greenhouses where "intermediate" temperature conditions are maintained (i.e., minimum nighttime temperature in the 16°C–18°C [61°F–64°F] range, roughly speaking).

*Rhynchostylis coelestis* is the species that has been most often used in the making of intergeneric hybrids with *Rhynchostylis*. The plant is short and stout, usually about 23–25 cm (9–10 in) tall, and its width is greater than its height. The species is widely distributed in Thailand, mainly in deciduous forests in mountainous areas; however, the species also inhabits low elevations in the Prachuab region of southwestern Thailand. In addition to Thailand, it is found in all of Indochina—Laos, Cambodia, and Vietnam.

The inflorescences of *Rhynchostylis coelestis* plants frequently bear 40–50 round, flat flowers measuring about 2 cm (0.75 in) across. They are well arranged in cylindrical fashion around a 20–30 cm (8–12 in) inflorescence. Flowering begins above the foliage. Two flowering spikes are common. The plant blooms in the second quarter of the year. As on the other two *Rhynchostylis* species, the flowers all open at the same time, and are delightfully fragrant. Their color usually is near-white, often suffused with violet, but the tips of the sepals and petals, together with the rather prominent lip, are deep violet on good clones. Especially prized is a form with deep pink markings. An *alba* form is found in northern Thailand (see Plate 77).

The availability of three rather distinct color forms enhances the potential of *Rhynchostylis coelestis* as a parent for intergeneric hybrids, because it widens the range of colors attainable in the offspring. When the violet form is combined with purple ascocendas, for example, the color transmitted by the latter is deepened and brightened. When the *alba* form is employed together with non-blue or non-violet ascocendas, a range of colors can be achieved; for this reason, many hybridizers have been attracted to the *alba* form.

A drawback of *Rhynchostylis coelestis* is that its flowers last only one or two weeks. Another shortcoming of the species, and one that often is transmitted to its *Vascostylis* descendants, is that its flowers have a somewhat nodding posture. While the pedicels of *Rhy. coelestis* issue forth from the flower stalk at an attractive upward angle, they generally "hook downward" at their outer ends where they join the flowers, so that the flowers tilt and are therefore harder to view.

*Rhynchostylis gigantea* is found mainly in northern Thailand, but its range extends into northern Burma and Indochina; it also is known from Borneo. The species blooms once yearly, in January or February. The inflorescence is 20–30 cm (8–12 in) long and carries about 50 flowers measuring about 3–4 cm (1.25–1.5 in) across. The flowers normally have white sepals and petals, with reddish purple blotches and spots, and a striking red-purple lip. A concolor wine-red variety, known as the Sagarik strain (Plate 78), was developed by Rapee Sagarik of Thailand from two deeply pigmented jungle plants. A prized all-white *alba* form also exists.

The plant often issues several inflorescences at the same time. This latter trait, along with the fact that nearly all the flowers open simultaneously, produces a very impressive display that is further enhanced by the heady fragrance of the flowers.

By crossing plants of the Sagarik strain with an *alba* form of the species, flowers with spots and blotches larger and darker than the typical type have been achieved. These never fail to attract attention; they are very eye-catching indeed. (See Plate 79.)

A form of *Rhynchostylis gigantea* endemic to the Philippines is subspecies *violacea* [*Rhy. gigantea* (Lindl.) ssp. *violacea* (Lindl.) E. A. Christenson]. It resembles

*Rhy. gigantea* florally in most respects but *Rhy. retusa* vegetatively. Some taxonomists treat it as a distinct species, either as *Rhy. violacea* or as *Anota violacea,* but Christenson (1986a, 169) concludes that the difference in the morphology of the flowers is not sufficient to warrant status as a separate species. It blooms from June through September in the Philippines and is uncommon in horticulture.

*Rhynchostylis retusa* is found in Thailand; it also occurs in southern and northern India, Burma, northern Malaysia, Indochina, the Philippines, Java, Borneo, and Sri Lanka. Its extensive range embraces warm lowland environments and, more often, somewhat montane habitats.

Few orchids excite viewers as much as a *Rhynchostylis retusa* blooming to its full potential. The flowers of *Rhy. retusa* are fragrant and all open at the same time. They last up to two weeks, a relatively short time by the standards of the cut-flower trade. The individual flowers are smaller than those of *Rhy. gigantea;* they measure about 1.3–2 cm (0.5–0.75 in) across; however, they are carried on much longer flower stalks. On well-grown plants, the raceme can reach 40 cm (16 in) in length and can have as many as 150 flowers. The plants often issue many flower shoots at blooming time. The coloring of the flowers is similar to that of *Rhy. gigantea* except that the purple spots are minute. A pure white strain also exists. In Thailand, the plants bloom in April and May, which is later than the flowering season of *Rhy. gigantea.* In the Philippines, however, the blooming season is June through September.

The two major combinations involving *Vanda* and *Rhynchostylis* are *Rhynchovanda* (*Rhynchostylis* × *Vanda*) and *Vascostylis* (*Ascocentrum* × *Rhynchostylis* × *Vanda*).

### *Rhynchovanda* (*Rhynchostylis* × *Vanda*)

The primary objective of *Rhynchovanda* breeding is to improve some of the traits of conventional *Vanda* hybrids: to increase the number and improve the arrangement of the flowers on their racemes; to elongate the inflorescence; to achieve simultaneous opening of the flowers; to shorten the height of the plant; and to do all of the foregoing without significant loss of flower shape or color or longevity, although some reduction of size of the flower is to be expected. This tall order rarely is fully achieved.

The most probable expectancy of flower number when a cross is made is the geometric mean of the number typically borne by each of the two parents on each blooming, not the arithmetic mean. Given the wide disparity between the very large number of flowers typically borne on an inflorescence of any of the *Rhynchostylis* species and the much smaller number borne by a typical *Vanda* hybrid, *Rhynchovanda* offspring will have a probable flower count closer to that of the

*Vanda* parent. For example, if a *Rhy. gigantea* with 50 flowers is crossed with a *Vanda* with 12 flowers, the geometric mean is 24 (the square root of [50 × 12]); the arithmetic mean is 31 ([50 + 12] ÷ 2). A flower count of 24 is typical of a cross of either *Rhy. coelestis* or *Rhy. gigantea* with a *Vanda* hybrid—31 would be high (but not unachievable).

The range of colors seen in *Rhynchovanda* hybrids thus far has been limited, with red-to-violet predominating (but see Plate 80). The color range may well be attributable to the choice of parents rather than being inescapable.

From the beginning in 1961 up to year-end 1992, a total of 70 *Rhynchovanda* crosses were registered. *Rhynchostylis coelestis* was the predominant *Rhynchostylis* species employed as a direct parent—31 times, as compared with 17 for *Rhy. gigantea.*

The first *Rhynchovanda* registered was *Rhv.* Blue Angel, which was to become an influential progenitor. It was registered by Francis Takakura of Hawaii in 1961. Its parents were *Vanda* Rothschildiana and *Rhynchostylis coelestis.*

In the first decade of *Rhynchovanda* breeding, a total of 22 hybrids were registered, about half of them by Hawaiian hybridizers—pioneers whose names appear repeatedly on the scroll of orchid history: Kirsch, Miyamoto, Mizuta, Moir, and Takakura. The contributions of breeders in Thailand and Singapore also were substantial. The use of *Rhynchostylis* species as direct parents was divided between *Rhy. coelestis* (11 times) and *Rhy. gigantea* (7 times). *Rhynchostylis retusa,* somewhat surprisingly, did not appear at all in that period; it first appeared as a parent in the 1980s, and only twice. *Rhynchostylis violacea* (more properly, *Rhy. gigantea* ssp. *violacea*) was recorded as a parent only once in the 1961–70 period. It was not recorded as a parent after that (through 1992).

Only the first recorded *Rhynchovanda, Rhv.* Blue Angel (*Vanda* Rothschildiana × *Rhynchostylis coelestis*) subsequently appeared as a *Rhynchovanda* parent to any significant extent; it was recorded on 5 occasions, and it has produced some highly awarded progeny.

While a number of *Vanda* species have been mated with *Rhynchostylis,* the dominant choice, by far, of hybridizers has been large-flowered standard *Vanda* hybrids, most notably hybrids with blue or purple flowers. Two well-known *Rhynchovanda* grexes are *Rhv.* Wong Yoke Sim (*V.* Rothschildiana × *Rhv.* Blue Angel) and *Rhv.* Galen Kanayama (*V.* Hilo Blue × *Rhv.* Blue Angel). They are favored for their indigo-purple color and for their erect, many-flowered inflorescences. Both also are old crosses; the first was registered in 1967, from Malaysia, and the second, in 1969, from Hawaii. They are still held in high regard.

Despite the modest size of its flowers, another of the better-known *Rhynchovanda* hybrids is *Rhv.* Sagarik Wine (*Vanda denisoniana* × *Rhynchostylis gi-*

*gantea*). It is known for its long, arching inflorescences bearing flat, well-distributed, vibrant dark red flowers; they open uniformly along the length of the raceme to make a very pleasing display.

### *Vascostylis* (*Ascocentrum* × *Rhynochostylis* × *Vanda*)

*Vascostylis* plants rather closely resemble ascocendas in floral appearance, but the flower spike of *Vascostylis* hybrids generally is longer and bears more flowers. The flowers also are more evenly distributed on the stalk, and there is less tendency toward crowding; however, in hybrids with a *Rhynchostylis coelestis* parent, the flowers often have a nodding carriage, which is considered a fault because it impairs viewing of the faces of the blooms. The plants are compact and no larger than ascocendas with a comparable amount of *Vanda* in their ancestry.

The man-made genus *Vascostylis* combines *Ascocentrum* with the two genera present in *Rhynchovanda*. Because there may be any order and frequency of combination of the three genera, there can be great variety in the results, as is the case with all other intergeneric combinations. For example, if a *Vascostylis* is produced by first making a *Rhynchovanda* and then crossing it with an *Ascocenda*, the number of flowers borne by the *Vascostylis* progeny is likely to be less than if the *Vascostylis* combination had been created by breeding a *Rhynchostylis* with an *Ascocenda*. In the first instance, *Vanda* parents, presumably with fewer flowers than either the *Rhynchostylis* or *Ascocentrum* parents, would have appeared in the lineage of the *Vascostylis* hybrid at least twice (at least once in the ancestry of the *Rhynchovanda* parent and at least once in the ancestry of the *Ascocenda* parent of the *Vascostylis*). Since one of the main objectives of *Vascostylis* breeding has been to increase the number of flowers typical of an *Ascocenda*, hybridizers overwhelmingly have chosen the latter route (i.e., crossing a *Rhynchostylis* with an *Ascocenda*).

The first *Vascostylis* registered was *Vasco*. Blue Fairy (*Ascocenda* Meda Arnold × *Rhynchostylis coelestis*); it was originated by Francis Takakura of Hawaii and registered in 1963, two years after he had registered the first *Rhynchovanda*. By year-end 1992, a total of 112 *Vascostylis* of strap-leaf *Vanda* ancestry had been recorded, of which 33 had a *Rhynchostylis* species as a direct parent. Twenty-five of these were *Rhy. coelestis*, 7 were *Rhy. gigantea*, and only one was *Rhy. retusa*.

Of the 112 total *Vascostylis* registrations, 80 had an *Ascocenda* as a direct parent, while only 26 had a *Rhynchovanda*; few had two *Vascostylis* parents (see Plate 81 for an exception). The desire to maximize the probability of progeny with a large number of flowers influenced the preference. Another reason was to keep down the size of the plants. The most frequently used *Rhynchovanda* parent was *Rhv.* Wong Yoke Sim (*Vanda* Rothschildiana × *Rhv.* Blue Angel). See Plate 82 for a *Vascostylis* with a *Rhynchovanda* parent.

Considering how many *Vascostylis* have been registered and awarded, it is surprising how little has been published about this trigeneric combination; 84 awards were published in the period 1970–92, by far the most bestowed on any of the intergeneric combinations listed in Appendix D, and 24 grexes were represented.

The advancement of *Vascostylis* breeding was spearheaded by Hawaiian hybridizers, among whom the names of Richard and Stella Mizuta and Susan and Robert Perreira frequently appeared. The best ascocendas were selected as mates: *Ascocenda* Meda Arnold, *Ascda.* Ophelia, *Ascda.* Yip Sum Wah, *Ascda.* Elieen Beauty, and *Ascda.* Tan Chai Beng. In the 1970s, Thai breeders, among others, also registered many crosses.

Despite all the activity and awards, the popularity of *Vascostylis* hybrids has been somewhat limited. One reason may be the one that restrains the supply of many intergeneric combinations; namely, low fertility, which makes it difficult to breed and produce many different grexes. Another reason may be the somewhat limited range of colors achieved thus far.

A review of the background and characteristics of the most recently awarded *Vascostylis* hybrids sheds light on some of these issues. In the 1985–92 volumes of the *Awards Quarterly* of the American Orchid Society, 36 *Vascostylis* cultivars were reported to have received flower-quality awards. Those awards covered only 7 individual grexes. Thirteen of the awards were given to cultivars in a single grex, *Vascostylis* Cynthia Alonso (*Vasco.* Doty × *Ascocenda* Yip Sum Wah). (See Plates 83 and 84). *Vascostylis* Doty, in turn, is a combination of *Ascda.* Medasand and *Rhynchostylis coelestis.* Both of these *Vascostylis* hybrids were registered by T. Orchids, a Thai nursery. Twelve of the awarded cultivars belonged to grex *Vasco.* Precious, a cross of *Rhy. coelestis* and *Ascda.* Yip Sum Wah, registered by Richard Mizuta of Hawaii in 1969. Thus, the two most heavily awarded hybrids of the 1985–92 period, *Vasco.* Cynthia Alonso and *Vasco.* Precious, together accounted for nearly 70% of the total awards, and both had *Ascda.* Yip Sum Wah as a direct parent. Another awarded hybrid, *Vasco.* Blue Velvet, also had *Ascda.* Yip Sum Wah as a parent. Altogether, therefore, *Ascda.* Yip Sum Wah sired 72% of the awarded plants listed in the *Awards Quarterly* in the 1985–92 period. The prodigious record of *Ascda.* Yip Sum Wah as a stud never ceases to amaze.

Despite the dominating presence of *Ascocenda* Yip Sum Wah, an orange-red-flowered grex, the majority of the plants awarded in the 1985–92 period displayed purplish flowers. Next in frequency were flowers of yellowish orange, burnt orange, or persimmon colors, which is not surprising. One hybrid had flowers described as "oxblood-red." Many had some tessellation, though usually not pronounced. With a greater variety of *Ascocenda* (and *Vanda*) parents, it should be possible to obtain a wider range of colors, and especially more shades of pink, along with greater contrasts of color. Plate 85 shows the attractive results that can be obtained by using parents having a strong representation of yellow-flowered vandas in their ancestry. Another goal for hybridizers is to produce larger flowers, while at the same time retaining the long, erect, and floriferous racemes characteristic of the best of the current *Vascostylis* hybrids.

In addition, the common fault of nodding flower carriage inherited from *Rhynchostylis coelestis* needs to be addressed. One Malaysian breeder, Yip Sum Wah, asserts that a very few *Rhy. coelestis* clones exist that do not transmit this trait, which is encouraging. He referred specifically to a collected plant of the *alba* form of *Rhy. coelestis* he has in his collection.

### *Perreiraara* (*Aerides* × *Rhynchostylis* × *Vanda*)

Another interesting combination incorporating *Rhynchostylis* is the man-made genus *Perreiraara*. It combines both *Aerides* and *Vanda* with *Rhynchostylis*. The first grex in this genus to be registered was *Prra*. Porchina Blue, by Robert Perreira of Hawaii in 1969. It was a cross of *Rhynchovanda* Blue Angel by *"Aerides" mitrata*. However, since this species properly belongs in the genus *Seidenfadenia* and not in the genus *Aerides,* the inclusion of this grex in *Perreiraara* is incorrect from a botanical point of view, even though it is placed there by the Registrar of orchid hybrids. The first botanically correct *Perreiraara* hybrid was *Prra*. Blue Charm, also by Robert Perreira and registered in 1980. It combines an *Aeridovanda* hybrid with a *Rhy. coelestis*. *Aeridovanda* plants also have been bred with *Rhy. gigantea* and *Rhy. retusa* to form new lines of breeding in the *Perreiraara* genus. It is too early to assess the future of *Perreiraara* hybrids with much confidence, but on the basis of *Prra*. Luke Thai (*Aerdv*. Vieng Ping × *Rhy. coelestis*), shown in Plate 86, they appear to have much promise, especially in terms of flower count, length of the inflorescence, and arrangement of the flowers.

### *Ronnyara* (*Aerides* × *Ascocentrum* × *Rhynchostylis* × *Vanda*)

Next in progression comes the genus *Ronnyara*, which combines the genus *Ascocentrum* with the genera included in *Perreiraara*. The first recorded *Ronnyara* hybrid was *Rnya*. Ronny Low, originated by Lum Hon of Kuala Lumpur, Malaysia, and registered in 1984. It is a cross of *Christieara* Virginia Braga by *Rhynchostylis gigantea*. In the same year, *Rnya*. Don-Ron Twin was registered; it combined *Rnya*. Ronny Low with an *Ascocenda*. In 1992, Michael Coronado of Florida registered *Rnya*. Melba Coronado, a cross of *Prra*. Luke Thai (Plate 86) with *Ascocenda* Fuchs Golden Nugget. In the same year, two other grexes were registered in the genus *Ronnyara*, by Suphachadiwong Orchids of Thailand; however, because each of these has as a parent *"Aerides" mitrata,* a species that properly belongs in the genus *Seidenfadenia,* they have not been counted in Appendix D.

Given the genera included in their lineage, *Ronnyara* hybrids should be capable of producing some fine results, even though, as of mid-1994, no plants in the genus had received an AOS award for flower quality.

## SOME COMBINATIONS WITH *RENANTHERA*

The genus *Renanthera* Loureiro is closely related to *Arachnis* and *Vandopsis*. Some of the species produce the showiest inflorescences to be found in the orchid kingdom. While the individual flowers are not large, and their sepals and petals are narrow, their scarlet, crimson, or yellow-orange colors are so electric, their flowers so numerous and so well displayed on their panicled inflorescences, that the overall effect is awesome. The aim of hybridizers who cross *Renanthera* with *Vanda* and *Ascocenda* has been to combine those qualities of *Renanthera* with the best traits of *Vanda* and *Ascocenda*. For example, both *Renantanda* (*Renanthera* × *Vanda*) and *Kagawara* (*Ascocentrum* × *Renanthera* × *Vanda*) have multiple-branched inflorescences, while, at the same time, the plants are shorter than most *Renanthera* species, and the sepals and petals of the flowers are wider.

The genus *Renanthera* contains some 10 or 12 species. It is distributed from the Himalayas to Burma, Malaysia, southern China, Thailand, Indochina, the Philippines, and southward to Indonesia and New Guinea. Only about 4 or 5 of the species have been used to any significant extent for hybridizing, both within the genus itself and in combination with other genera such as *Vanda* and *Ascocenda*. Most *Renanthera* are quite sensitive to cold.

The foremost species in the genus is *Renanthera storiei* Reichb. *f.*, which is endemic to the Philippines. It is a very tall plant; it can extend to more than 4 m (12 ft) in height if properly supported. It bears luminous scarlet or crimson flowers heavily mottled with darker red blotches, especially on the lateral sepals. The flowers are borne in great profusion on long, many-branched panicles that can reach out as far as 1 m (3 ft). The individual flowers measure 5–9 cm (2–3.75 in) vertically and about 3.5–5 cm (1.5–2 in) horizontally. The plants thrive in full sun, and must have it if the full color potential is to be achieved. In the Philippines, *Ren. storiei* blooms throughout the year—an additional reason for its popularity with hybridizers. See Plate 87 for a *Ren. Storiei* hybrid.

*Renanthera coccinea* Loureiro is a vinelike species indigenous to Thailand, southern China, and Indochina. The flowers and flowering habit resemble those of *Ren. storiei*, but the lateral sepals are somewhat brighter red. The inflorescences may carry more than 75 flowers. Individual flowers measure 4.5–5 cm (1.75–2 in) across and about 6 cm (2.5 in) vertically. Peak blooming is in the spring, but the plants can bloom sporadically throughout the year.

*Renanthera imschootiana* Rolfe, a species found in Assam (India) and Indo-china, is considerably shorter than *Ren. storiei* and *Ren. coccinea,* and so are its branched inflorescences. Even so, the inflorescence, which has short internodes, can reach more than 0.5 m (1.5 ft) in length and bear a multitude of blooms; an awarded plant was described as having "many buds and 506 flowers on two spikes." The relatively broad lateral sepals appear scarlet at a distance but actually are yellow-orange heavily overlaid with fine red spots; the petals and dorsal sepal are orange-yellow. The petals are spotted with darker red. The flowers measure about 3.3–5 cm (1.3–2 in) horizontally and 5–6 cm (2–2.5 in) vertically. *Renan-thera imschootiana* blooms mostly in the summer months. It can bloom on very small plants.

*Renanthera monachica* Ames is a relatively small species endemic to the Philip-pines. It is a tidy plant. The stems of the plant can reach 60 cm (2 ft) in height. The inflorescences are simple or, more usually, branched, and measure nearly 30 cm (1 ft) in length. The flowers are star-shaped and are somewhat loosely arranged on the inflorescences; the number of flowers typically is more than 20. The color of the flowers is muted orange-yellow, heavily and evenly dotted with scarlet spots. (See Plate 88.) The individual flowers measure about 2.5–3 cm (1–1.25 in) across and approximately the same vertically. *Renanthera monachica* plants bloom mainly from January to June.

*Renanthera philippinensis* Ames & Quisumbing is another *Renanthera* species endemic to the Philippines. The plants are smaller than *Ren. storiei* and about 20 cm (1 ft) taller than *Ren. monachica.* Originally, *Ren. philippinensis* was regarded as a smaller variety of *Ren. storiei.* The inflorescences have many-flowered pani-cles, like *Ren. storiei.* The flowers are crimson or scarlet-red and measure 1.9 cm (0.75 in) across and 2.5–3.2 cm (1–1.25 in) vertically. The lip has a white center, an attractive feature that generally is transmitted to hybrid offspring (e.g., *Ren.* Manila [*Ren. philippinensis* × *Ren.* Brookie Chandler]). Among the good qualities of *Ren. philippinensis,* in addition to moderate height, are short internodes, simi-lar to those of *Ren. imschootiana,* and short leaves. The plants are capable of flow-ering throughout the year, and they thrive in full sun.

### Renantanda (*Renanthera* × *Vanda*)

*Renantanda* hybrids fall between *Renanthera* and *Vanda* in plant size, flower shape, and flower number, depending on which species are used, and the frequency of each. Flower color is strongly influenced by the *Renanthera* heritage; it usually ranges from crimson red through orange to orange-yellow; however, some of the brilliance of color of *Renanthera* flowers generally is lost in the *Renantanda* hy-brids, and fading also has been a problem, although more lately this has largely been corrected by careful selection of the *Ren. storiei* parent when that species is involved.

*Renantanda* hybrids have been spectacularly successful as a product for the cut-flower industry of Southeast Asia. The very showy, long-lasting inflorescences are excellent for display and interior decorating. For comparable brilliance of red-or-ange color on more moderately-sized plants, most hobby growers outside of the tropics may prefer one of the smaller-growing intrageneric *Renanthera* hybrids that bloom when the plants are quite small. An excellent one is *Ren.* King Crim-son, a cross of *Ren. philippinensis* and *Ren. monachica* (Plate 89). Another is *Ren.* Memoria Marie Killian (Plate 87), which combines genes from *Ren. imschootiana*, *Ren. monachica,* and *Ren. storiei.* Perhaps equally attractive would be a *Kagawara* combination (*Ascocentrum* × *Renanthera* × *Vanda*). See Plate 90.

Up to the end of 1992, a total of 170 *Renantanda* hybrids had been recorded. In nearly four-fifths of the cases, a *Renanthera* species served as a direct parent, and *Ren. storiei* accounted for about two-thirds of these. Two things are clear: first, compact size was not a consideration in most of the efforts, and, second, the year-long flowering habit and large flower sprays of *Ren. storiei* were highly re-garded attributes desired in the intergeneric hybrid offspring. *Renanthera coccinea* shares the same characteristic of large, bright-colored flowers and long, branched sprays, but it generally blooms more sporadically during most of the year. It was the second most frequently used species; however, it was a direct parent only 18 times—a sharp drop from the popularity of *Ren. storiei. Renanthera imschootiana* served as a parent only 10 times. *Renanthera philippinensis* was used only 4 times, a surprisingly weak record considering the desirably modest size of the plant, the bright color of its flowers, which are borne in profusion on many-paniceled branches, and its habit of blooming throughout the year. Its lateral sepals are somewhat narrower than those of some of the other *Renanthera* species, but when crossed with *Vanda* hybrids with considerable *V. sanderiana* ancestry, the breadth of the latter's sepals and petals should counteract most of this tendency. The ex-planation for its lack of use may lie in the fact that *Ren. Philippinensis* was not an easily available species until relatively recently.

*Renanthera monachica,* a species to which one would expect hybridizers inter-ested in the hobbyist market to turn, was a direct parent on 12 occasions when *Renantanda* hybrids were being created. Unfortunately, *Ren. monachica* does not have the scarlet or crimson color saturation of the other species that have been used. Apparently, this consideration caused hybridizers to lose interest in this species; between 1971 and 1992, only 2 *Renantanda* hybrids were recorded with *Ren. monachica* as a parent. *Renanthera monachica* did find a prominent niche for itself, however, in the breeding of *Renanthopsis* (*Renanthera* × *Phalaenopsis*). A famous example is the beautiful *Rnthps.* Mildred Jameson, originated by Henry M. Wallbrunn of Florida and registered in 1969; it inspired the breeding of a number of other *Renanthopsis* hybrids using *Renanthera monachica.*

### *Kagawara* (*Ascocentrum* × *Renanthera* × *Vanda*)

*Kagawara* hybrids combine the vegetative and floral features of the trigeneric parentage. The combination is not easy to make, but the results can be very rewarding (see Plate 90). The seedlings grow rapidly and the plants flower when quite young—two traits not normally associated with complex intergeneric crosses. A major objective of breeders of *Kagawara* has been to produce plants of less size than *Renantanda.* Preservation of the brilliance of *Renanthera*—oftentimes diminished in the transition to *Renantanda*—has been another objective, along with inflorescences that are easier to pack for overseas shipment. The addition of *Ascocentrum* genes helps to reinforce flower roundness and flatness, traits also inherited from the customary *Vanda* ancestry (i.e., vandas with a *V. sanderiana–V. coerulea* background).

Depending on the ancestral species, *Kagawara* hybrids have provided plants of interest either for the cut-flower trade or for the hobby grower, or for both. To date, however, most breeding has been directed with the cut-flower industry in mind, but it is to be hoped that more breeding will be done on behalf of the amateur greenhouse grower. One Florida hybridizer, Claire Sidran, has been exploring the potential of this line of breeding, as well as other intergeneric combinations with *Renanthera,* such as *Lutherara* (*Phalaenopsis* [including *Paraphalaenopsis*] × *Renanthera* × *Rhynchostylis*).

The first two *Kagawara* hybrids were registered in 1969. One was *Kgw. William Doi, Jr.* (*Renanthera* Kilauea × *Ascocenda* Meda Arnold), originated by John Tew of Hawaii and registered by H. Kagawara, after whom this intergeneric combination was named. The other, *Kgw.* Firebird (*Ren. storiei* × *Ascda.* Red Gem), was originated and registered by Richard Mizuta, also of Hawaii. Shortly thereafter, *Kagawara* registrations appeared from Singapore and Thailand, as well as others from Hawaii.

By the end of 1992, a total of 64 *Kagawara* crosses had been registered. Of these, 61 had an *Ascocenda* parent; only 6 had a *Renantanda* parent. Seventeen had *Renanthera storiei* hybrids as a parent, while only 2 had a *Kagawara* parent, which lends support to the reports that *Kagawara* hybrids share the tendency of other Sarcanthinae trigenerics to be infertile.

Throughout the period, *Renanthera* species continued to be employed as direct parents of *Kagawara*—a total of 34 times. Again, by far the most popular species was *Ren. storiei;* it was a parent on 19 occasions, or more than half of the total number of *Kagawara* crosses that had a *Renanthera* species as a direct parent. *Renanthera philippinensis* was employed 9 times, or more than a quarter of the total usage of *Renanthera* species; this is in marked contrast with the pattern of *Renantanda* breeding, where its use was rare. *Renanthera monachica* was not used even once, probably because breeders did not wish to dilute the bright red color of *Ren. storiei* and *Ren. philippinensis;* moreover, for cut-flower growers, the smaller plant size of *Ren. monachica* carried no special advantage. *Renanthera coccinea,* the second most popular *Renanthera* species parent of *Renantanda,* appeared only twice as a parent of a *Kagawara* hybrid, perhaps because *Ren. storiei* was considered superior in most relevant respects.

*Kagawara* breeding, as the record seems to confirm, has leaned toward emphasizing suitability for the cut-flower trade and for display purposes, although this need not have been the case, because *Kagawara* plants can bloom when quite small; even mature plants can be accommodated in hobby greenhouses. To bloom well, the plants do require somewhat brighter light than vandas, and they should be hung high. They are free blooming and very cheerful looking, with their brilliant red or orange colors. Although difficult to breed and propagate, they are very satisfying plants once established. Much further breeding lies ahead, as the potential becomes more widely known to hobby growers in the United States and elsewhere.

### *Holttumara* (*Arachnis* × *Renanthera* × *Vanda*)

This genus produces plants with good-sized, showy, red-orange flowers in profusion on long, branched inflorescences. Flower shape is influenced heavily by the *Arachnis* and *Renanthera* heritage. An example is shown in Plate 91. A considerable number of *Holttumara* grexes have been made for the flower trade of Southeast Asia. The plants are large and require bright sun and high temperatures in order to grow and bloom well. Consequently, they are not good choices for greenhouse growers in temperate climates.

## SOME COMBINATIONS WITH
## *PARAPHALAENOPSIS* AND *PHALAENOPSIS*

Prior to 1964, *Paraphalaenopsis* and *Phalaenopsis* were lumped together in the genus *Phalaenopsis* Blume. In that year, A. D. Hawkes established *Paraphalaenopsis* as a separate genus, and botanists now widely accept it as such; *Sander's List of Orchid Hybrids,* however, still does not distinguish between them for registration of hybrids. Sweet (1980, 118), in his monograph on *Phalaenopsis,* states:

> *Paraphalaenopsis* plants are not compatible genetically with those of *Phalaenopsis,* but they will hybridize with members of other genera, *Aerides, Arachnis, Euanthe, Luisa, Papilionanthe, Renanthera* and *Vanda.*

He regards that fact as being of critical importance in defining the limits of the genus *Phalaenopsis*.

Because the role of *Phalaenopsis* in intergeneric combinations with *Vanda* and other Sarcanthinae species has been very limited, and because so much has been published about that genus, this section will concentrate on combinations with *Paraphalaenopsis* species, with only passing reference to *Phalaenopsis,* as needed,

The genus *Paraphalaenopsis* A. D. Hawkes currently contains four species: *P. denevei* J. J. Smith; *P. labukensis* Lamb, Chan & Shim; *P. laycockii* (M. R. Henderson) A. D. Hawkes; and *P. serpentilingua* (J. J. Smith) A. D. Hawkes. All the *Paraphalaenopsis* species are endemic to Borneo. They have long, terete leaves, quite unlike any of the true *Phalaenopsis* species. The inflorescences are stout and stiff. Only *P. serpentilingua* has racemose inflorescences; the others are umbellate. The inflorescences normally bear 5–7 clustered flowers, but on very strong, mature plants, the number may range up to 15. The flowers of all of the species are fragrant. When *Paraphalaenopsis* plants are bred to vandas or ascocendas, the plants tend to inherit these floral characteristics and they resemble semi-terete vandas vegetatively.

*Paraphalaenopsis* species must be grown wet. Failure to provide almost constant wetness is the principal reason for their demise under cultivation. The need for high moisture tends to be transmitted to some of their intergeneric hybrids, such as *Lutherara* (*Phalaenopsis* [including *Paraphalaenopsis*] × *Renanthera* × *Rhynchostylis*).

*Paraphalaenopsis denevei* flowers (Plate 92) are an attractive greenish yellow color and their margins are undulating. The clearly demarcated midlobe of the lip is dark red, and the prominent side lobes are red on their upright apices; basally they are white with minute red dots. The species can flower several times each year, and plants may issue several inflorescences on each occasion, with up to 15 flowers on each inflorescence. The flowers measure about 5 cm (2 in) across.

*Paraphalaenopsis labukensis* flowers give an overall impression of reddish brown and are quite spectacular. Actually, the background color is a rather light greenish yellow, which shows clearly around the edges of the sepals and petals. Heavily overlaid on the rest of the surface of the sepals and petals is a rich reddish brown, covered with a multitude of tiny greenish yellow spots. The lip has a similar combination of colors, but the spotting is more prominent. The lateral sepals are relatively wide and flare forward and sideways.

*Paraphalaenopsis laycockii* has an umbellate inflorescence that bears 9–15 flowers. The flowers are light mauve or rose and are about 7.5 cm (3 in) across, somewhat larger than those of *P. denevei* and *P. sepentilingua*. On its midlobe apex, the lip is colored like the sepals and petals, but it becomes a brownish red basally and on the side lobes.

*Paraphalaenopsis serpentilingua* has yellowish white flowers with recurved and twisted sepals and petals. The inflorescences are longer than those of the other species (up to 35 cm [14 in] long), and the flowers are smaller than those of *P. denevei* and *P. laycockii*. The yellow lip is the most prominent feature: its midlobe resembles a snake's forked tongue.

While *Paraphalaenopsis* species cannot be crossed directly with *Phalaenopsis,* the species in the genus can be crossed with one another. Two such intrageneric hybrids are *P.* Boediardjo (*P. denevei* × *P. laycockii*) and *P.* Sunny (*P. serpentilingua* × *P. denevei*). *Paraphalaenopsis* Sunny was bred to *Phal.* Doris, to produce the intergeneric hybrid *"Phalaenopsis"* Doris Thornton. Given the incompatibility of the individual *Paraphalaenopsis* species with *Phalaenopsis,* it is odd that two of those *Paraphalaenopsis* species in combination could be successfully combined with *Phalaenopsis.* *"Phalaenopsis"* Doris Thornton is unique; no other such cross has been recorded. One cannot help but wonder whether something strange happened during meiosis or during fertilization (e.g., a roaming insect may have thwarted the hybridizer).

## *Vandaenopsis* with *Paraphalaenopsis* Parentage

Combining *Vanda* with *Paraphalaenopsis* species has produced some interesting, free-blooming plants with fine-colored flowers of good substance and oftentimes with fragrance as well. Flower shape has not been round or flat, and the crosses may properly be called novelty hybrids. The leaves of the plants are semi-terete. Plant size, as well as floral traits, are much influenced by the particular species ancestry of the *Vanda* parent.

The first recorded *Vandaenopsis* with *Paraphalaenopsis* parentage was *Vdnps.* Jawaii (*Vanda sanderiana* × *P. denevei*), registered in 1938 by Frank Atherton of Hawaii. It is surprising that this line of experimentation was attempted so early, because, up to that time, very little hybridizing had been undertaken even *within* the *Vanda* genus. It is another example of the pioneering spirit of the early Hawaiian orchid growers. In the following three years, 3 more hybrids were made in what now is Indonesia; *V. coerulea, V. insignis,* and *V. luzonica,* respectively, were the *Vanda* parents, and in each case, *P. denevei* was the other parent.

Up to the end of 1960, 11 *Vandaenopsis* hybrids had been made between strap-leaf vandas and *Paraphalaenopsis* species, all of the latter being *P. denevei.* During the next 10 years, 31 more hybrids were registered, reflecting considerable interest in this line of breeding during that decade. Interest diminished somewhat during the following two decades. From 1971 to 1992, 22 hybrids were recorded, including some with *P. laycockii.* By the end of 1992, a total of 64 *Vandaenopsis* hybrids combining strap-leaf *Vanda* and *Paraphalaenopsis* parentage had been registered. The great majority were created in Singapore, but a number were made elsewhere. The novelty of the crosses has had great appeal.

Two outstanding *Vandaenopsis* hybrids with *Paraphalaenopsis* parentage are

*Vdnps.* Twinkle (*Vanda* Kekaseh × *P. laycockii*) and *Vdnps.* Laycock Child (*V.* Mandai Amber × *P. laycockii*). A description of each will help to convey what this line of breeding can produce.

*Vandaenopsis* Twinkle was registered by Singapore Orchids in 1982. The hybrid can show noticeable variation in color, but it usually has cheery, canary-yellow flowers, with somewhat narrow sepals and petals that are covered with very minute red dots arranged in a linear pattern longitudinally. The lip is bright red. The substance of the flowers is good. The *Vanda* parent of *Vdnps.* Twinkle, *V.* Kekaseh, has *V. cristata* and *V. insignis* as its parents. Prior to producing *Vndps.* Twinkle, *V.* Kekaseh had been used by Singapore Orchids to make a number of *Vanda* hybrids in the 1974–82 period, and also to make some crosses with *Ascocenda* and *Renanthera* plants. Clearly, it was well regarded as a parent.

*Vandaenopsis* Laycock Child also was registered by Singapore Orchids, in 1990. It has eye-catching, round, flat, rose-pink or yellowish flowers (there can be considerable variation). The lip may be yellow or reddish. The *Vanda* Mandai Amber parent combines genes from *V. coerulea, V. dearei, V. liouvillei,* and *V. sanderiana*—a veritable potpouri.

*Vandaenopsis* made with *Paraphalaenopsis,* like the latter species, need to be kept moist if they are to thrive.

## *Vandaenopsis* with *Phalaenopsis* Parentage

*Vandaenopsis* hybrids of the type combining strap-leaf *Vanda* plants with *Phalaenopsis* have been designed to produce flowers that extend the boundaries of *Phalaenopsis* blooms. The results depend on the chosen combinations of *Vanda* and *Phalaenopsis* species and hybrids.

Apparently the goals of the hybridizers have varied greatly. Evidence of this is found in the diversity of different species that has been tried. As of the end of 1992, 10 different *Vanda* species and 8 different *Phalaenopsis* species had been used to produce *Vandaenopsis* grexes. Consequently, *Vandaenopsis* hybrids display a great variety of flower counts, sizes, shapes, and colors.

The crosses combining large-flowered, standard *Vanda* hybrids with large-flowered, standard *Phalaenopsis* thus far have not produced progeny with flowers more attractive than those of the parents. The most interesting combinations have been ones with *V. cristata* parentage. One such grex is *Vandaenopsis* Revelation (*V. cristata* × *Phal.* Fairvale). (See Plate 20 for a photograph of *V. cristata.*) *Phalaenopsis* Fairvale has large, light pink flowers. *Vandaenopsis* Revelation 'Jacqueline', HCC/AOS, was described in the award description as having 18 flowers and buds on two spikes. The flowers had "unique coloration, deep purple-red overlaying a cream-green base; lip solid dark purple red; crystalline texture, heavy substance." The natural spread was 7.7 cm (3 in). Judging from the width-to-length ratios of the sepals and petals, the flowers were open or star-shaped.

The first *Vandaenopsis* hybrid of the type discussed in this section was recorded in 1931, a cross of *Phalaenopsis amabilis,* a species bearing many large, flat, white flowers, with *Vanda tricolor* var. *suavis* (see Plate 9). The hybridizer was M. Chassaing, the chief gardener of the Rothschild chateau Ferrieres-en-Brie, Seine & Marne, France. In the same year, Chassaing also was credited with making the most famous *Vanda* hybrid of all time, *V. Rothschildiana* (*V. coerulea* × *V. sanderiana*).

Through 1960, an additional 4 *Vandaenopsis* hybrids involving *Phalaenopsis* were recorded. In the following decade, 12 were registered, and from 1971 through 1992, another 23, making a grand total of 40—a respectable number but well short of the total of 64 *Vandaenopsis* with *Paraphalaenopsis* ancestry.

Ten *Vandaenopsis* cultivars representing 4 grexes made with *Phalaenopsis* parentage had received flower awards from the American Orchid Society by year-end 1992. Three of these 4 grexes had *Vanda cristata* as a parent, which clearly demonstrates that the greatest success in *Vanda-Phalaenopsis* combinations has been in producing novelty types of hybrids, rather than in attempts to produce better "standard" types of *Vanda* or *Phalaenopsis* hybrids.

### *Devereuxara*

(*Ascocentrum* × *Phalaenopsis* [including *Paraphalaenopsis*] × *Vanda*)

*Devereuxara* hybrids represent an additional step of complexity beyond *Vandaenopsis,* because they add *Ascocentrum* to the mix. *Devereuxara* crosses are very difficult to make, and it is equally difficult to generalize about them. As in the case of *Vandaenopsis,* so much depends on whether the *Phalaenopsis* ancestry refers to *Paraphalaenopsis* or to *Phalaenopsis.*

A total of 41 *Devereuxara* had been registered up to the end of 1992. The great majority had a *Phalaenopsis* hybrid as a parent or grandparent. For an example, see Plate 93. Only 9 of the 41 had any *Paraphalaenopsis* in their background. Only one of them, *Dvra.* Hawaiian Adventure (*P.* Boediardjo × *Dvra.* Anna Paul), had both *Paraphalaenopsis* and *Phalaenopsis* in its ancestry, but via an indirect route.

Generally, one parent of *Devereuxara* hybrids was an *Ascocenda;* in every instance, an *Ascocenda* appeared somewhere on the family tree, if not as a direct parent of the hybrid then as a grandparent or great-grandparent. The only time an *Ascocentrum* appeared in the lineage, other than in the prior forming of an *Ascocenda,* was as a parent of *Asconopsis* (*Asctm.* × *Phalaenopsis*) Irene Dobkin (*Phal.* Doris × *Asctm. miniatum*)*;* this trail-blazing *Asconopsis* was a parent of 6 of the 41 *Devereuxara* hybrids recorded up to the end of 1992. In 4 of these 6 cases, the other parent was an *Ascocenda.*

In the period 1985–92, 5 flower awards were bestowed on *Devereuxara* hybrids by the American Orchid Society. Three of the awarded plants were cultivars of *Dvra.* Hawaiian Delight (*Phalaenopsis* Barbara Moler × *Dvra.* Hawaiian Rainbow). *Devereuxara* Hawaiian Rainbow has a *Phalaenopsis* as one parent and, on the other side, another one as a grandparent, so the representation of *Phalaenopsis* is substantial—three out of four grandparents. The fourth grandparent is an *Ascocenda, Ascda.* Ruth Shave, (*Vanda* Eisensander × *Ascda.* Ophelia). The flowers of each of these three awarded plants showed little influence of the *Ascocenda* ancestry in shape, size, or even color, except, perhaps, for the yellow hue of their flowers and the linear spotting of lavender. Another awarded *Devereuxara* with comparable ancestry, *Dvra.* Edith Normoyle 'Norm's Dream', HCC/AOS (*Dvra.* Great Expectations × *Phal.* Norm's Fantasy), also appears little different from a standard *Phalaenopsis,* judging from the award description, except that its 30 flowers is a respectable improvement over its *Phalaenopsis* ancestry.

If *Devereuxara* Hawaiian Delight and *Dvra.* Edith Normoyle are typical of *Phalaenopsis-Ascocenda* combinations, the addition of *Ascocenda* genes seems scarcely worth the effort, given the fact that the hybrids are harder to make and to grow than are *Phalaenopsis* themselves.

It is more difficult to evaluate combinations with *Paraphalaenopsis* background because of the paucity of awarded plants, but at least some novelty would be anticipated, and that was true for the one plant that received an award in the 1985–92 period—*Devereuxara* Dreamer 'Baby Doll', AM/AOS, a cross of *Ascocenda* Yip Sum Wah and *P. laycockii.* When judged, the plant had 17 flowers of reasonable size (6.6 cm [2.6 in]) and a round outline, but the individual sepals and petals were about half-again as long as they were wide. The flowers were nicely arranged; their color was orange-pink, changing with age to rosy pink, which apparently charmed the judges.

Based on the very limited evidence available, and taking into account experience with the section of *Vandaenopsis* breeding that incorporates *Paraphalaenopsis,* it seems that those who wish to explore the potential attractions of *Devereuxara* and *Vandaenopsis* are more likely to be pleased with the outcome of plants with *Paraphalaenopsis* in their ancestry, rather than those made with standard *Phalaenopsis* hybrids.

### Vandewegheara

(*Ascocentrum* × *Doritis* × *Phalaenopsis* [including *Paraphalaenopsis*] × *Vanda*)

Thirteen of this very complex intergeneric hybrid had been registered by year-end 1992. The results have been mixed, even though two plants received HCC flower-quality awards from the American Orchid Society. The flowers of one of the plants (see Plate 94) very closely resemble those of a standard *Phalaenopsis* and would seem to have no advantage over countless straight *Phalaenopsis* hybrids. The other awarded *Vandewegheara, Vwga.* Tom Raso (*Ascda.* Yip Sum Wah × *Doritaenopsis* Red Coral), shows its *Ascocenda* heritage in its coloring and flower shape.

The first hybrid to be registered was *Vandewegheara* Jerry Vande Weghe, recorded in 1975 and named after the originator. In the 1980s, Helmut Paul of Hawaii registered a number of grexes. Thus far, none of the *Vandewegheara* grexes has achieved significant attention or popularity.

## SOME COMBINATIONS WITH *NEOFINETIA*

*Neofinetia* H. H. Hu is a monotypic genus consisting of *Neofinetia falcata* (Thunb.) H. H. Hu, a dwarf species of about 15 cm (6 in) in height, found in China, Korea, Japan, and the Ryukyu Islands. The plants issue many shoots from their base, making them rather bushy, but they are compact plants. They bloom mainly in the summer and early fall. The flowers are pure white and sweetly fragrant, especially at night. Pink-flowered clones of *Neof. falcata* exist, but they are rare in cultivation and practically unknown outside of Japan. The inflorescences are erect and bear 5–6 flowers measuring about 2.5 cm (1 in) horizontally; they are somewhat longer vertically, not counting the 5 cm (2 in) nectary spur. With the long spurs, the flowers look like small angraecums. See Plate 95. The plant and its intergeneric hybrids are much appreciated in Japan.

*Neofinetia falcata* plants do best under intermediate to cool conditions, and their roots should be kept moist. The species is an excellent choice for growing under lights or on windowsills.

Hybridizers are attracted by the dwarf size, the many basal shoots, and the very long pedicels and ovaries of this species. As a parent, *Neofinetia falcata* tends to open up crowded inflorescences when they exist on the other parent. Hybridizers have combined *Neof. falcata* with *Vanda* to make *Vandofinetia*; with *Ascocentrum* and *Vanda* to make *Nakamotoara* (Plate 96); with *Rhynchostylis* and *Vanda* to make *Yonezawaara*; with *Ascocentrum, Rynchostylis,* and *Vanda* to make *Darwinara* (Plate 97); and with *Aerides* and *Vanda* to make *Vandofinides*. These combinations are not often seen outside of Japan. With the exception of *Vandofinetia* Virgil (*Neof. falcata × V. cristata*), no AOS awards to any cultivar of any of these man-made genera had been published in the *Awards Quarterly* up to mid-year 1993. Judging from the award description and photograph, *Vf.* Virgil 'Botanicals', HCC/AOS, was not an improvement over either parent.

Also to be mentioned are *Ascofinetia* (*Ascocentrum × Neofinetia*) and *Neostylis* (*Neofinetia × Rhynchostylis*), neither of which contain any *Vanda* parentage and, strictly speaking, for that reason are outside the scope of this book.

Most of the innovative breeding of intergeneric combinations with *Neofinetia* has been taking place in Japan, and the work of O. N. Takaki has been especially notable. One *Darwinara, Dar.* Charm (*Neof. falcata × Vascostylis* Tham Yuen Hae), registered by Takaki in 1987, is a delightful miniature. The clone 'Blue Star' (Plate 97) has been mericloned and shown in the United States. It blooms frequently.

All of the *Neofinetia* combinations are recommended on a trial basis, especially for those who seek compact plants with vandaceous flowers that are out of the ordinary; however, flower quality will be highly variable, and it is premature to speculate about which line of breeding will produce the best results. The hybrids adapt well to growing under lights and on windowsills.

## SOME COMBINATIONS WITH *VANDOPSIS*

*Vandopsis* Pfitz. is closely related to *Arachnis* Blume and *Trichoglottis* Blume. The genus is found in Southeast Asia, southern China, the Philippines, and New Guinea. The genus contains more than 8 species, but only 4 have been of horticultural interest. Those 4 are *Vandopsis gigantea* (Lindl.) Pfitz.; *Vdps. lissochiloides* (Gaud.) Pfitz.; *Vdps. parishii* (Veitch & Reichb. *f.*) Pfitz., known best for its variety *marriottiana* Reichb. *f.* and nowadays usually placed in the genus *Hygrochilus* Pfitz.); and *Vdps. warocqueana,* now placed in the genus *Sarcanthopsis* Garay as *Sarcanthopsis* (Rolfe) Garay.

*Vandopsis* species have been of little interest to hobby growers in temperate climates. Neither have they been highly regarded in the cut-flower trade. Most of the species are difficult to grow and bloom. In the best of circumstances, they are not frequent bloomers. They have been of some value to hybridizers, however, who have combined them with a number of other genera and species in the Sarcanthinae subtribe; for example, with *Rhynchostylis* to form *Opsistylis.*

*Vandopsis gigantea* plants are widely distributed in Southeast Asia. They are about 50–75 cm (20–30 in) tall and have a thick vegetative stem. The leaves are 30 cm (12 in) long and about 6.4 cm (2.5 in) wide. The inflorescence is 25–35 cm (10–14 in) long, with an arching or drooping habit; it bears 7–15 round flowers measuring about 6.4 cm (2.5 in). The flowers have some fragrance and are long-lasting, due to their fleshy substance. The background color of the flowers is yellow, marked with reddish brown-purple blotches. (See Plate 98). The lip is white with lavender markings. In Thailand, the plant blooms from April to June. As of the end of 1992, *Vdps. gigantea* had received 9 AOS awards, mostly for culture rather than flower quality.

*Vandopsis lissochiloides* plants are found in Thailand, the Philippines, and Indonesia. The plants grow lithophytically or terrestially and are tall, often reaching 1 m (3 ft) in height; they have an erect inflorescence of 1–2 m (3–6 ft), bearing about 25 flowers that open several at a time. The flowers measure about 5 cm (2 in) and are fragrant. The sepals and petals are chartreuse with pronounced reddish brown spotting. The flowers are not at all flat, and their shape is more open than that of *Vdps. gigantea.*

*Vandopsis parishii* (now *Hygrochilus parishii*) is best known for its variety *marriottiana,* which is by far the most attractive member of the group. The plants grow epiphytically at fairly high elevations in Thailand, Burma, and Indochina.

Plate 49. *Ascocenda* Yip Sum Wah (*V.* Pukele × *Ascocentrum curvifolium*). No other *Ascocenda* approaches the performance of this 1965 hybrid, both in awards received and as a parent—not only of ascocendas but also of other intergeneric combinations. Photograph by Charles Marden Fitch.

Plate 50. *Ascocenda* John De Biase 'Lava Flow', FCC/AOS (*Vanda* Kasem's Delight × *Ascda.* Yip Sum Wah), reputedly a triploid made with a tetraploid *Ascda.* Yip Sum Wah. This grex is the most highly awarded *Ascocenda* since its great parent. Registered by R. F. Orchids in 1983. Photograph by Bob Smith.

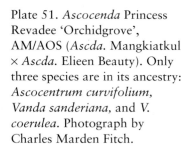

Plate 51. *Ascocenda* Princess Revadee 'Orchidgrove', AM/AOS (*Ascda.* Mangkiatkul × *Ascda.* Elieen Beauty). Only three species are in its ancestry: *Ascocentrum curvifolium*, *Vanda sanderiana*, and *V. coerulea*. Photograph by Charles Marden Fitch.

Plate 52. *Ascocenda* Memoria Emily Grove 'Thai Beauty', AM/AOS *(Vanda coerulea × Ascda.* Princess Revadee). The reintroduction of *V. coerulea* has produced larger, intensely colored purple flowers on long, erect inflorescences, but the flowers lack the flat, round dorsal sepal of the other parent. Photograph by Charles Marden Fitch.

Plate 53. *Vanda limbata × Ascocenda* Red Gem, an unregistered hybrid. *V. limbata* has seldom been used as a parent of *Vanda* or *Ascocenda* hybrids. The other parent is also unusual; its parents are *V. merrilli* (Plate 6) and *Ascocentrum curvifolium* (Plate 11). Photograph by Charles Marden Fitch.

Plate 54. *Ascocenda* Khun Nok (*Vanda lamellata × Ascda.* Madame Panni), an example of the fascinating results that can occur when ascocendas are bred back onto some of the *Vanda* species. The red on the inner half of the lateral sepals is inherited from the *V. lamellata* parent (see Plate 4). Photograph by Charles Marden Fitch.

Plate 55. *Ascocenda* Lek (*Vanda luzonica* × *Ascda.* Flambeau). *Ascda.* Flambeau is a cross of *Ascda.* Yip Sum Wah and *V. tessellata.* The combination has produced an intriguing pattern of colors. Photograph by Charles Marden Fitch.

Plate 56. *Ascocenda* Suk Sumran Beauty (*Vanda* Gordon Dillon × *Ascda.* Yip Sum Wah). The mating of these two outstanding hybrids has produced some excellent progeny, as illustrated by this vibrant red hybrid. Photograph by Charles Marden Fitch.

Plate 57. *Ascocenda* Memoria Arthur Freed (*Ascda.* Yip Sum Wah × *Vanda cristata*). The unusual lip reflects the influence of *V. cristata*, which is the main attraction of that *Vanda* species (see Plate 20). Photograph by Charles Marden Fitch.

Plate 58. *Ascocenda* Memoria Arthur Freed. Another cultivar of this hybrid, demonstrating both the variability in color and the wisdom of buying several seedlings of promising new crosses. Photograph by Charles Marden Fitch.

Plate 59. *Ascocenda* Laksi 'Red Ruby', AM/AOS (*Vanda* Thonglor × *Ascocentrum curvifolium*). *Vanda* Thonglor (*V.* Jennie Hashimoto × *V.* Diane Ogawa), registered in 1974, was one of the better *Vanda* hybrids created in Thailand in the 1965–75 period. Photograph courtesy of R. F. Orchids.

Plate 60. *Ascocenda* Tubtim Velvet 'White Angel' (*Ascda.* Jenny Donald × *Vanda* Kultana Gold). This bicolored white and yellow *Ascocenda* has been mericloned and is widely distributed. It is deservedly very popular. Photograph by Charles Marden Fitch.

Plate 61. *Ascocenda* Bangyikhan Gold 'Viboon' (*Ascda.* Yip Sum Wah × *Vanda* Rasri). *Vanda* Rasri was an important step forward toward good yellow *Vanda* flowers. Its originator, Amnuay Sathira-sut, is well known for his yellow vandas and line-breeding of *Ascda.* Yip Sum Wah. This vibrant orange *Ascocenda* combined both efforts. Photograph by Charles Marden Fitch.

Plate 62. *Ascocenda* Duang Porn 'Orchidgrove', AM/AOS (*Vanda* Thananchai × *Ascda.* Yip Sum Wah); another combination of a famous yellow *Vanda* with *Ascda.* Yip Sum Wah, to produce a good orange *Ascocenda*. Photograph by Charles Marden Fitch.

Plate 63. *Ascocenda* Fuchs Butter Baby (*Ascda.* Fuchs Golden Nugget × *Ascocentrum curvifolium*). An alternative approach to producing orange-yellow ascocendas is to use genes from *Asctm. miniatum*, which *Ascda.* Fuchs Golden Nugget possesses; the further introduction of *Asctm. curvifolium* reinforces brightness of color. Photograph by Charles Marden Fitch.

Plate 64. *"Vanda"* Nellie Morley ([*Papilionanthe teres* × *V. tricolor* = *"V."* Emma Van Deventer] × *Vanda sanderiana*) has some tetraploid forms that have been much used in Hawaii for cut flowers. It has been called "the pride of Hawaii," and is very popular there. Photograph by Charles Marden Fitch.

Plate 65. *Arachnis* Maggie Oei (*Arach. flos-aeris* × *Arach. hookeriana*), is a cornerstone of the Southeast Asia cut-flower industry. Note the scorpionlike appearance. Photograph by Charles Marden Fitch.

Plate 66. *Aranda* Christine (*Arachnis hookeriana* × *Vanda* Hilo Blue). In nurseries in the tropics, *Aranda* and *Arachnis* usually are grown in rows on the ground. *Aranda* Christine is one of the most famous *Aranda* hybrids both in its own right and as a parent of other hybrids. Photograph by Charles Marden Fitch.

Plate 67. *Aranda* Nancy (*Arachnis hookeriana* × *Vanda dearei*). The complete absence of visible spots on the flowers of this hybrid presumably is the result of using the *luteola* form of *Arachnis hookeriana*, which has no spots. Photograph by Charles Marden Fitch.

Plate 68. *Aranda* Eric Mekie (*Aranda* Lucy Laycock × *Vanda luzonica*). The inflorescences display the appeal of this hybrid to the cut-flower trade. Photograph by Charles Marden Fitch.

Plate 69. *Mokara* Redland Sunset 'Robert's Ruby', HCC/AOS (*Aranda* Singapura × *Ascocenda* Yip Sum Wah), a vibrant *Mokara* originated by Kultana Orchids in Bangkok. Photograph by Bob Smith.

Plate 70. *Aerides multiflora* is one of the most floriferous species in the *Aerides* genus. Photograph by Charles Marden Fitch.

Plate 71. *Aeridovanda* Fuchs Cream Puff (*Vanda* Charles Goodfellow × *Aerides lawrenceae*). The round shape and flatness, and the pleasing contrast of the bright red lip against the white background, illustrate the charm of aeridovandas with *Aerides lawrenceae* parentage.

Plate 72. *Aeridovanda* Kinnaree (*Vanda* Teoh Chee Keat × *Aerides lawrenceae*). Perfect arrangement and spacing of round, flat, and beautifully colored flowers, but their posture is rather nodding. Photograph by Charles Marden Fitch.

Plate 73. *Aeridovanda* Arnold Sanchez (*Aeridovanda* Vieng Ping × *Aerides lawrenceae*). The bright yellow petals and sepals, the pink midlobe of the lip, and the prominent nectary all combine to produce striking flowers. Photograph courtesy of R. F. Orchids.

Plate 74. *Christieara* Rene Gerber (*Ascocenda* Bonanza × *Aerides lawrenceae*), an eye-catching, red-speckled, orange flower. Photograph by Charles Marden Fitch.

Plate 75. *Christieara* Fuchs Confetti (*Ascocenda* Theptong × *Aerides lawrenceae*). This grex and cultivar display an attractive blending of colors, which is one of the charms of *Christieara* hybrids. The prominent lip is a feature of *Christieara* plants. Photograph by Charles Marden Fitch.

Plate 76. *Lewisara* Fatimah Alsagoff 'Zahrah'. The individual flowers and the inflorescences of this complex intergeneric hybrid combine pleasing attributes of its *Aerides, Ascocentrum, Arachnis,* and hybrid *Vanda* ancestry. Photograph by William Smiles.

Plate 77. *Rhynchostylis coelestis* f. *alba*. The *alba* form has been a favorite of hybridizers when breeding for *Vascostylis* (*Ascocentrum* × *Rhynchostylis* × *Vanda*) hybrids. The nodding carriage typical of the flowers of this species is evident here, but the erect inflorescence and attractive flowers are very good traits. Photograph by Charles Marden Fitch.

Plate 78. *Rhynchostylis gigantea*, Sagarik's strain. By crossing two of the rare red mutations of *Rhy. gigantea*, Rapee Sagarik produced a brilliant red color form that breeds true to color. Photograph by Charles Marden Fitch.

Plate 79. *Rhynchostylis gigantea*, in this instance a man-made cross combining an *alba* form of the species with one of the Sagarik strain. Photograph by Charles Marden Fitch.

Plate 80. *Vanda* Pissamai × *Rhynchostylis gigantea*, an unregistered *Rhynchovanda* hybrid with fragrant flowers. *Vanda* Pissamai is a cross of *V.* Thananchai, a rust-speckled, tawny yellow *Vanda*, by *V. laotica* (= *V. lilacina*), a seldom-used species. Photograph by Charles Marden Fitch.

Plate 81. *Vascostylis* Five Friendships (*Vasco.* Seng × *Vasco.* Prapin), a cross of two *Vascostylis* (*Ascocentrum* × *Rhynchostylis* × *Vanda*) hybrids, which is not the usual route of hybridizers; here a *Rhynchostylis coelestis* is both a maternal and paternal grandparent. Photograph by Charles Marden Fitch.

Plate 82. *Vascostylis* Charles Marden Fitch (*Rhynchovanda* Flamingo × *Ascocenda* Theptong). *Ascda.* Theptong has *Vanda denisoniana* as a parent, and that parent's influence on flower color and shape is visible in this hybrid. Photograph by Charles Marden Fitch.

Plate 83. *Vascostylis* Cynthia Alonso 'Bridget', HCC/AOS (*Vasco.* Doty × *Ascocenda* Yip Sum Wah). A stunning hybrid combining genes from *Rhynchostylis coelestis* and from two of the foremost red ascocendas, *Ascda.* Medasand and *Ascda.* Yip Sum Wah. Photograph courtesy of R. F. Orchids.

Plate 84. *Vascostylis* Doty (*Ascocenda* Medasand × *Rhynchostylis coelestis*). This is the flower arrangement, number, and shape hybridizers are seeking in *Vascostylis* hybrids, along with good color. The slight cupping of the flowers can easily be forgiven. Photograph by Charles Marden Fitch.

Plate 85. *Vascostylis* Nong Kham (*Ascocenda* Jiraprapa × *Vasco.* Prapin). The beautiful color combination is similar to that of *Ascda.* Tubtim Velvet (Plate 60). Both parents have *Vanda* Kultana Gold, a concolor yellow *Vanda*, in their ancestry. Photograph by Charles Marden Fitch.

Plate 86. *Perreiraara* Luke Thai 'Pat Howell', HCC/AOS (*Aeridovanda* Vieng Ping × *Rhynchostylis coelestis*), a complex and intriguing combination, but very few have been registered or made. Photograph by Charles Marden Fitch.

Plate 87. *Renanthera* Memoria Marie Killian 'Eric's Red Imp', CCM/AOS, (*Ren.* Merritt Island × *Ren.* John Tew). Most of the progenitors of this floriferous hybrid are *Ren. storiei*, but *Ren. imshootiana* and the much smaller-growing *Ren. monachica* also are represented. Photograph by Charles Marden Fitch.

Plate 88. *Renanthera monachica* 'Suzanne'. This clone is an exceptionally well-colored and well-branched specimen. Photograph by Charles Marden Fitch.

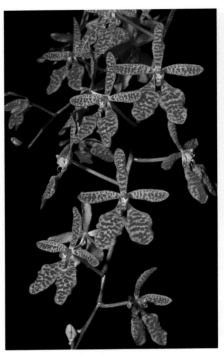

Plate 90. *Kagawara* Yoon Weng-Low 'TOF' (*Renanthera philippinensis × Ascocenda* Yip Sum Wah). *Ren. philippinensis* is a considerably shorter species than *Ren. storiei*, but it has most of the latter's fine qualities. This *Kagawara* combines those with the powerful genes of *Ascda*. Yip Sum Wah. Photograph by Charles Marden Fitch.

Plate 89. *Renanthera* King Crimson *(Ren. philippinensis × Ren. monachica)*, a hybrid of two relatively compact species. The wide, flat lateral sepals, combined with the clear, bright markings, make this a good prospect for breeding with *Ascocenda*. Photograph by owner William Smiles.

Plate 91. *Holttumara* Park Nadesan (*Aranda* Hilda Galistan × *Renanthera storiei*), an intergeneric combination specifically made for the floral trade. The *Renanthera* parent increases the floriferousness of the progeny. Photograph by Charles Marden Fitch.

Plate 92. *Paraphalaenopsis denevei*. Both the flowers and the growing habit of this terete-leaved species are well displayed, and are illustrative of the genus. Photograph by Charles Marden Fitch.

Plate 93. *Devereuxara* Hawaiian Exotic (*Phalaenopsis* Donna Shiro × *Dvra.* Hawaiian Sunset). The *Phalaenopsis* parent seems to have controlled the appearance of the flowers, which is typical of this type of *Devereuxara* hybrid. Photograph by Charles Marden Fitch.

Plate 94. *Vandewegheara* Hawaiian Flare (*Phalaenopsis* Puna Gold × *Vwga.* Mauna Kea). *Vandewegheara* hybrids combine *Ascocentrum, Doritis, Phalaenopsis* and *Vanda*. Since the flowers generally seem indistinguishable from those of *Phalaenopsis*, one must wonder what the advantage is of the complexity of the breeding. Photograph by Charles Marden Fitch.

Plate 95. *Neofinetia falcata*. This dwarf species is a favorite in Japan, a native habitat, because of the compactness and free-blooming habit of the plants, as well as the charm and sweet fragrance of the *Angraecum*-like flowers. Photograph by Charles Marden Fitch.

Plate 96. *Nakamotoara* Cinderella 'Hawaii' (*Vandofinetia* Premier × *Ascocentrum curvifolium*). The *Vanda* ancestor is *V. lamellata* (see Plates 4 and 21). *Neofinetia* combinations such as this are very popular in Japan. Although not yet widely grown in the United States and Europe, they probably soon will be. Photograph by Charles Marden Fitch.

Plate 97. *Darwinara* Charm 'Blue Star', HCC/AOS (*Neofinetia falcata* × *Vascostylis* Tham Uyen Hae), a well-named hybrid made by O. N. Takaki of Japan, is a freely blooming dwarf plant ideal for growing under lights or on a windowsill. Photograph by Richard Clark, courtesy of Cal-Orchid, Inc.

Plate 98. *Vandopsis gigantea*. This species has very heavy, leathery flowers. In combination with *Vanda* and *Ascocenda*, it has not lived up to hybridizers' expectations. Photograph by Charles Marden Fitch.

Plate 99. A well-maintained Thai orchid nursery, displaying how vandas and ascocendas are grown in Thailand. The ground is covered with sand, for cleanliness and to provide humidity through evaporation after the daily watering. Wood laths provide 50% shading.

They are considered difficult to grow and bloom in cultivation in the tropical low-lands; they do best in an environment similar to that of the native habitat of *Vanda coerulea* (Kamemoto & Sagarik 77; also see 173 for pictures of *Vdps. parishii* and the variety *marriottiana*). The plants have a short, stout stem. The leaves are broad and fleshy, like those of *Phalaenopsis*. The inflorescences are almost 30 cm (1 ft) long, nearly erect, and bear 6–12 flowers arranged in two rows. The flowers are round, of heavy substance and waxy texture, and measure about 5 cm (2 in) across. The sepals and petals of the variety *marriottiana* are a bronzy red or laven-der on their outer portion, shading to a lighter lavender hue toward their center, and white near their base. The lip varies from lavender-purple to greenish purple. The species blooms from March to May; the flowers last up to two weeks.

*Vandopsis warocqueana* (now *Sarcanthopsis warocqueana*) is a very tall litho-phytic plant native to New Guinea. The stout, stiff, branched inflorescences are about 25 cm (10 in) long and bear many flowers 2.5 cm (1 in) wide. The flowers are greatly cupped. Their color is creamy yellow, irregularly overlaid with fine dots of brownish red. The lip has the light yellow background color of the sepals and petals. (See Teoh, 163, for a fine picture of the species.) Many of the flowers have a nodding posture which displays the much heavier reddish coloring of the back of the dorsal sepal.

### Opsisanda (*Vandopsis* × *Vanda*)

When combined with *Vanda*, *Vandopsis* contributes only two traits of value—in-creased flower substance and, generally, yellowish color attractively overlaid with small magenta or maroon spots. The good traits of the *Vanda* parents, for the most part, either are not improved or are worsened in the *Opsisanda* progeny. The best *Opsisanda* hybrids, by far, are these made with *Vandopsis* (now *Hygrochilus*) *parishii* var. *marriottiana*.

Through the end of 1992, 53 *Opsisanda* hybrids had been registered: 27 with *Vandopsis lissochiloides*; 12 with *Vdps.* (now *Hygrochilus*) *parishii*; 11 with *Vdps. gigantea*; and 3 with *Vdps.* (now *Sarcanthopsis*) *warocqueana*. Only 8 *Opsisanda* cultivars, representing 4 grexes, had received AOS awards as of year-end 1992, and all but 2 of those 8 awards were bestowed prior to 1966. The two more re-cently awarded plants (one in 1978 and one in 1985) each received the minimum number of points needed to obtain a flower award (i.e., 75 points, out of a possi-ble score of 100).

### Wilkinsara (*Ascocentrum* × *Vanda* × *Vandopsis*)

The goal of this line of breeding has been to produce compact plants with round, waxy flowers, an increased number of flowers, and improved arrangement. By concentrating on *Vandopsis* (now *Hygrochilus*) *parishii* (almost invariably the variety *marriottiana*, which has the roundest petals and sepals and the best coloring of the *Vandopsis* species), the odds of producing something good were maximized.

Of the 15 *Wilkinsara* hybrids registered prior to 1993, 10 had *Vandopsis* (now *Hygrochilus*) *parishii* as a direct parent. The only other *Vandopsis* parent was *Vdps. gigantea,* and that was used only once.

Judging from awards granted to *Wilkinsara,* the goals of the hybridizers sometimes were attained. When awarded, *Wknsra.* Redland Sunrise 'Robert', HCC/AOS, a cross of *Vandopsis* (now *Hygrochilus*) *parishii* with *Ascocenda* Chaisiri, had 21 flowers and 6 buds on its inflorescence. The flowers measured 4.1 cm (1.6 in) across. The color of the blossoms was tawny tangerine with clear white centers. The lip was darker orange in tone. The flowers were waxy, very flat, well proportioned, and evenly spaced on a horizontal inflorescence—altogether a very appealing combination.

Another awarded *Wilkinsara* is Golden Delite 'Cynthia', AM/AOS, a cross of *Vandopsis* Sagarik (*Vdps. parishii* × *Vdps. gigantea*) and *Ascocenda* Yip Sum Wah. When awarded, it had 7 flowers and 8 buds on a long, erect inflorescence. Flower color was rich gold covered with fine, brick-red spots. The flowers had very heavy substance. They were nicely spaced on their spike but were not very well arranged around it. The individual blooms measured 6.4 cm (2.5 in) across.

The record, although limited, suggests that *Wilkinsara* crosses are much to be preferred to *Opsisanda.*

## CONCLUSION

Only a fraction of the intergeneric combinations that can be made with the genus *Vanda* have been described, mainly those that have received the most attention from hybridizers and their customers. The amount of past and present attention, however, is not necessarily a reliable prediction of future potential. It is highly probable that some of the sparingly tried intergeneric combinations have yet to disclose their full potential, and it is at least equally possible that some of the more frequently attempted combinations may yield even better results in the future than they have up to now.

We are only beginning to see the extraordinary effects that can come from using tetraploid (instead of diploid) forms of species and hybrids as stud plants. Although many of the intergeneric and multigeneric vandaceous crosses produce small populations, tissue culture can be employed to make superior clones widely available at reasonable cost.

# PART TWO

# Plant Selection and Care

# 7

# A Guide to Buying Seedlings

## GENERAL CONSIDERATIONS

The typical buyer of plants in the *Vanda* alliance (i.e., vandas and other members of the Sarcanthinae subtribe) has little choice but to buy unbloomed seedlings much of the time. Relatively few mericlones (reproductions made by tissue culture) are available, unlike the situation with the *Cattleya* alliance, although, increasingly, more and more are being produced each year. Progress in improving the quality of hybrids in the *Vanda* alliance, especially vandas and ascocendas, was so fast until the early 1990s, breeders say, that most mericlones would have been passé by the time they reached the age of blooming. By the mid-1990s, however, that was no longer true; the very best cultivars will be very difficult to surpass in the future, given the very high level of flower quality already attained.

Neither are divisions of outstanding plants available, with an occasional exception. Fine plants in bloom sometimes are offered for sale, but if they are really outstanding, the seller justifiably will charge a high price. Unless very lucky, he had to bloom a large number of plants in order to obtain a few choice ones.

All this means that success in buying plants is often a matter of chance. Until the offspring of a cross begin to bloom, no one can predict with any certainty how early a typical cultivar will bloom or how good the flowers will be. Given the marked differences in the genetic traits of different grexes and the wide range of quality within most grexes, you the potential buyer must do whatever you can to tilt the odds in your favor. That requires that you know what to look for in making your selections.

Three considerations are fundamental: (1) the size or age of a plant, (2) the condition of the plant, and (3) its ancestry.

## Seedling Size and Age

When assessing seedling size, a distinction needs to be made between differences in the size of sibling seedlings of a particular cross and differences of plant size that reflect the various ages of plants of different origin. Most orchid nurseries offer plants in several age groups, but for any given cross, in only one age group. How should you choose among the possibilities?

In selecting plants from a batch of siblings of the same grex, it generally is best to choose the largest plants. When ordering from a catalog, if more than one sibling size of a cross is offered, specify the largest size, even though the price is higher. The largest siblings have displayed their greater vigor. That vigor will persist, and the plants will be easier to grow. They are likely to bloom at an early age and to have larger and more numerous flowers than their more laggard siblings. There is no certainty of this, of course, but the odds very definitely lie in that direction. Plants that bloom at an early age are very likely to be free-blooming throughout their life—an observation sometimes called Mehlquist's Law, after Gustav A. L. Mehlquist, a famed orchidologist and horticulturist who has contributed enormously to our scientific knowledge of many kinds of plants.

The largest available size usually is referred to as blooming size or "in flower." A blooming-size plant, regardless of its physical size, can be expected to flower in the upcoming blooming season of the particular cross. Almost always, the term "blooming size" means that the plants shipped have not yet produced a mature inflorescence—most sellers know that buyers do not want to buy the grower's culls; however, previously bloomed mericlones are acceptable because mericlones produce uniformity of bloom (except where mutations occur). Nonmericloned plants in bloom usually are sold to florists or to walk-in customers.

The next largest size is called near-blooming size. They often are siblings of the blooming-size plants of the same cross, but have grown more slowly. These plants, if well grown by the customer, should be sufficiently mature to flower in a year's time, or perhaps slightly longer.

For younger plants, catalogs usually state the average height or leaf-spread of the plants, or the width of the basket or pot, rather than an estimate of the amount of time before blooming age.

Moving farther down the age ladder, when infant seedlings are removed from flasks, they are placed in what are called community pots (often abbreviated to "compots"). Most community pots contain 15–35 very small seedlings. The seedlings stay in the community pot until they have enough roots to enable the individual plants to adapt to a basket or pot of their own. How long that will take depends on the genes and the culture provided. Some nurseries sell compots or plantlets recently removed from the pots, and flasks also sometimes are offered for sale at the retail level, but the choices are very limited, because demand is small.

Most hobby growers will find it advantageous to spend their money on blooming-size or near-blooming size plants (the former if both sizes are available for the same grex). The higher initial cost of the larger plants is less than the cost of heat, fertilizer, labor, and pots needed to grow small seedlings to maturity. There also is the ever-present possibility of losing young seedlings; they are quite susceptible to poor culture and disease.

Thai and other Southeast Asian growers have so perfected the art of growing vandaceous orchids that plants of blooming size often are produced within two years from removal from community pots. Commercial growers in Florida and Hawaii ordinarily cannot quite match this, and it takes most amateurs in the United States or Europe five or six years to reach the same stage. Most hobby growers will be better off having a collection of fewer plants, but of flowering size, than to devote much space to plants that will not provide pleasure from their flowers for some years to come (and may then prove to be of inferior quality).

### Assessing the Condition of a Seedling

When you visit a nursery, buy the sturdiest plants you can find. Never accept weak or diseased plants, either at a nursery or through the mail. Leaves should show no signs of bacterial or fungal spotting. Even though a past infection seems to be cured, there may be some residual infection in the vascular tissue. In particular, avoid plants with rough-feeling, dark purple or black elongated areas of scar tissue on some of the leaves (usually the lower ones). These are symptoms of a fungus disease called *Phyllosticta* leaf spot. (For more about this disease, see Chapter 9.)

If you have the opportunity to do so, carefully examine the root structure. Look for numerous long, plump roots with pronounced new tips. Branched roots are better than unbranched ones, and newly emerging roots just above the plant's base, or higher up, are very desirable. If you must choose between a somewhat taller plant with less vigorous roots and a somewhat shorter one with obviously superior roots, make the better roots the deciding factor. The plant will make a faster transition to new conditions.

Avoid plants that are in rotting wooden baskets or have badly disintegrated potting medium around the roots. Such conditions promote growth of fungus, bacteria, and insects. Ask nurseries either to send plants in clean baskets or pots, and in fresh potting medium if any is used, or to ship the plants bare-root, with strong, live roots. If those conditions cannot be satisfied, and you cannot resist the cross or find it elsewhere, clean the plants immediately upon arrival and put them in new containers with fresh medium (if you use any). If you do not do so, you may bring infections and insects into your growing quarters, and they may spread to the rest of your collection. A little algae on wooden baskets, by itself, is not a cause for concern; it can readily be corrected by dipping the basket into a solution of an algicide (see Chapter 9).

## ANCESTRY OF SEEDLINGS

It is very difficult to predict the quality and characteristics of vandaceous hybrid seedlings with any degree of reliability. The ancestry of the plant determines such genetic traits as flower color, shape, size, and number; flower arrangement on the inflorescence; age of first blooming; and maximum frequency of blooming. Culture and maturity, however, determine whether genetic potential is realized.

To improve your chances of getting good plants, buy from knowledgeable orchid sellers. Investigate the nurseries that offer the plants you are considering. Some vendors are also hybridizers; some are distinguished orchid judges, or are widely traveled in Southeast Asia, where they have seen hundreds of different crosses. If you buy from such nursery owners, they will be pleased to share their expectations about a cross with you. They had to form an opinion before they bred or purchased the seedlings they are offering. Less-knowledgeable sellers all too often wind up with disappointing crosses other hybridizers have unloaded on them at bargain prices. You may spend a little more, but you will benefit by buying from those who know quality and know which crosses are likely to produce it. As Rodney Wilcox Jones, the patriarch of the orchid community in the early decades after World War II, liked to say, "Good quality remains long after the price has been forgotten."

Look for a rich diversity of species ancestry when buying standard *Vanda* or *Ascocenda* hybrids, and keep in mind that the proportions should be right. These considerations are not so important when buying novelty types. For *Vanda* crosses, there should be substantial representation of *V. sanderiana* and *V. coerulea* ancestry and a small representation of *V. dearei, V. luzonica,* and *V. tricolor.* Not all of the last three *Vanda* species are needed; any two of them will suffice, but without two of those three, the odds for promising genetic diversity, especially with respect to flower colors and patterns, will be considerably reduced. For ascocendas, *Ascocenda* Yip Sum Wah is a highly desirable parent or grandparent.

For vandas with solid yellow or chartreuse flowers, look for seedlings with a predominant amount of *Vanda sanderiana* ancestry (preferably the *alba* form), combined with a *V. dearei* great-grandparent. The seedlings may also have some *V. coerulea, V. luzonica,* and *V. tricolor* ancestors, provided they are remote; in combination, they contribute greater frequency of blooming, longer inflorescences, a larger number of flowers, and tolerance of a wider range of temperatures.

Those contributions can be preserved even after the influence of those species on flower color has been removed through generations of selective breeding.

Ascocenda hybrids with an *Ascocentrum* direct parent have rather small flowers—generally with a natural spread of 4.5 cm (1.75 in) or less; the larger the ratio of *Vanda sanderiana* and *V. coerulea* grandparents and great-grandparents to those of *Ascocentrum* origin, the larger the expected flower size.

The nearer the appearance of *Ascocentrum curvifolium* in the lineage of an *Ascocenda,* the greater the likelihood of red or orange flowers, usually with fine spots on the sepals and petals. The nearer the appearance of *Asctm. miniatum,* the smaller will be the flowers and the greater the likelihood of yellow or orange flowers.

For vandas, ascocendas, and even more complex intergeneric combinations of species, *Vanda coerulea* imparts blue coloration if it serves as a direct parent. If it serves as one of the grandparents, the likelihood of blue or purple flowers is strong but by no means assured—red and pink flowers often result.

The most vibrant pink vandas and ascocendas generally result from crossing a plant with deep red flowers, with no brownish tones, with one with clear dark purple or deep blue flowers. (The vast majority of the offspring, however, will be a dull purple, with grayish tones.) A hybrid made with two pink-flowering parents produces offspring with paler pink flowers, or flowers of some other color. Hybrids made with one bright red parent and one pink parent are most likely to produce pink-flowering progeny of satisfactory quality.

Combinations with *Vanda sanderiana* as a direct parent are likely to bloom at a later age than combinations in which *V. sanderiana* is a more remote ancestor. Multiplicity of species ancestry increases the probability that some plants of the cross will bloom at an early age and frequently, and offset the contrary effects of *V. sanderiana.*

## Polyploidy

Polyploidy may favorably affect petal width, overall flower size, substance, and intensity of color. Polyploids do not always produce superior plants, by any means, but, when purchasing seedlings, consider the potential superiority of polyploids, especially triploids. More triploids are becoming available, such as the initial making of *Ascocenda* John De Biase (*Vanda* Kasem's Delight × *Ascda.* Yip Sum Wah). Its *Ascocenda* parent was a tetraploid. (Remakes may exist that did not use a tetraploid parent.) My recommendation is to buy triploids and tetraploids whenever possible, because progress is increasingly likely to come from greater use of polyploid parents. Diploid hybrids, however, should not be excluded; refinements still are being made through remakes of earlier crosses, using superior parents, and noteworthy advances are resulting from greater richness in the mix of ancestry of new diploid hybrids. See Chapter 2 for more on polyploidy.

### Remakes

Many commercial and amateur growers believe that remakes of an outstanding cross have a high probability of producing offspring of equal or superior quality, and that the favorable odds are even higher if the original parents are used, provided that they are of high quality. That proposition is not confirmed by experience—the remakes may be better or worse than the original cross. It is rather like rolling a pair of dice; if the first roll produces two sixes, that has no predictive value for the next roll. (See Chapter 2 for a discussion of the distribution of genes.)

An interesting confirmation on a more specific level comes from Roy Fukumura (1986). He remade *Ascocenda* Yip Sum Wah and *Ascda*. Peggy Foo each three times, and *Ascda*. Meda Arnold twice, using the original parents He says that none of the remakes approached the excellence of quality of the original cross. Consequently, he no longer remakes crosses.

Fukumura's "original" *Ascocenda* Meda Arnold, however, actually was a remake of a cross first made by Sideris of Hawaii in 1950, but using different cultivars as parents. Fukumura's remake turned out to be far superior to the original cross. It was his subsequent attempts to equal or exceed his first very successful effort that failed.

The proper conclusion is that remakes may be better or worse than the original cross, but the better the offspring of that original effort the more unlikely it is that repetitions will be still better. The better the first roll of the dice, the less the probability of exceeding it on the following throw.

Instead of using the identical parents (or superior ones) to remake a cross, breeders sometimes make sibling crosses of the best cultivars of the original cross. (Those retain the parental grex name: *Vanda* Jones 'Willy' × *V*. Jones 'Johnnie' = *V*. Jones.) Occasionally, some tetraploid plants will be produced from sibling crosses, or from selfings (a cultivar crossed with itself). An outstandingly successful example is *Ascocenda* Yip Sum Wah; the best tetraploid plants in that grex were created by sibling crosses and selfings, which then were used as parents of such famous grexes as *Ascda*. John De Biase. Do not hesitate to buy seedlings produced by sibling crosses in outstanding grexes, especially if the siblings chosen were superior cultivars. Quality, on the whole, begets quality, even if there is no guarantee.

# 8

# Providing the Right Environment

## ENVIRONMENT VERSUS CULTURAL PRACTICES

Most orchid hobbyists spend far too much time inquiring about specific cultural practices, such as how often to water, what kind of fertilizer to use, what dosage to apply, and so forth. They seem to believe that the best growers have some guarded secrets that, if shared, would enable them to achieve equally outstanding results. Let it be said here and now that there are no secrets. Good culture, for the most part, is not about what a grower does to his or her plants; it is about the kind of environment he or she provides for them.

The environment is determined by the combined interactions of light, heat, moisture, and movement of air. The problems and opportunities created by these environmental factors and their interactions vary considerably from one locality to another, and from one season of the year to another. Growers must respond to these changing conditions in a timely fashion, as best they can, if they are to achieve the potential of their plants.

Most species and hybrids in the *Vanda* alliance are quite adaptable, but they are more tolerant of some suboptimal climatic conditions than they are of others. Few growers have the time or money or capability to optimize every single variable every day of the year, despite the availability of electronic sensors, controls, and artificial lighting. Even in the wild, Mother Nature does not provide ideal conditions for any species much of the time.

This chapter presents acceptable ranges for the most important climatic factors. Try to keep climatic conditions within these ranges throughout the year, to the extent that external circumstances permit. Those circumstances, and the ability to modify them, are not identical in all geographic locations, and they are seasonal. The challenge is to learn how best to make whatever adjustments are practicable, and to learn when to make them, because timing is all-important.

143

## The Importance of Balance

Our main concern as growers is to ascertain and supply the *balanced* environment our plants need in order to thrive, given the diverse and changing conditions with which each of us must contend. Everything else is secondary to this goal. Specific cultural practices are good, bad, or indifferent only as far as they influence the total balance of the conditions under which a plant is trying to grow. Plants are genetically programmed to grow and bloom if given half a chance to do so. Given that opportunity—and that is the grower's responsibility—they will develop the abundant root structure on which the growth of the rest of the plant depends.

Optimal conditions for a plant are not static or independent of one another. For one thing, they change as the plant grows; a plant's requirements as a small seedling are different from its needs at maturity. Conditions and needs also change with the time of day and the season. Furthermore, the optimal amount or timing of any single climatic element, such as humidity, depends on other climatic elements, such as temperature and amount of air circulation. In an arid climate, powerful air movement can be very desiccating; in highly humid locations, substantial movement of air is needed to prevent bacterial and fungal infections. High humidity in the greenhouse is beneficial when the day is sunny and hot, but it can be harmful if the day is cold and damp outside.

It is the combination and balance of the climatic conditions, in conjunction with the supplementary cultural practices (such as fertilizing and potting), that unite to form a good or poor total environment. A continual balancing act is called for.


## Greenhouse Microclimates

Regardless of geographical location, there is no such thing as a single climate inside a greenhouse when it is filled with plants. Every greenhouse has many microclimates. The greatest difference between highly successful orchid growers and mediocre ones is that the former constantly pay careful attention to the condition of each of their plants and to its microclimate, while the latter do not. While many differences in microclimates are not critical to a plant's survival over a period of a year or two, they may affect the frequency of blooming and the quality of its flowers. Orchids grow slowly and die slowly.

Microclimates are greatly influenced by the seasonal fluctuations in the climate outside; periodic adjustments of shading, equipment, and placement of plants must be made within the greenhouse to counterbalance those fluctuations. The greater the seasonality of the exterior climate, the greater and the more frequent must be the adjustments. Improving the microclimate of one plant often affects the climate of nearby plants; to mention only two factors, it may alter the amount of light some plants receive, or the air currents around them.

An ideal microclimate for one plant may be less than ideal for another of the same genus, because of genetic differences. An ideal climate for a *Vanda coerulea*, for example, is not ideal for a *V. sanderiana*, although growers in most zones can grow both species quite well by finding an appropriate microclimate for each, because each tolerates some departure from the ideal. As another example, *V. dearei* is especially sensitive to cold, and some of its hybrid descendants may inherit this trait; clearly, such plants should be given the warmest location in the greenhouse. Yet another example: in the higher latitudes, *Arachnis* and *Papilionanthe* species and their hybrids require almost all the light and heat they can get. That means that they should be given the most light and warmth available; otherwise they will seldom bloom. *Ascocentrum ampullaceum* and *Asctm. curvifolium,* on the other hand, require some coolness to initiate blooming, and if need be should be moved to a cool location in the greenhouse about a month or so before their normal blooming period.

## BASIC ENVIRONMENTAL REQUIREMENTS

Vandas and their intergeneric hybrids require warmth, with some spread between the daytime and nighttime temperatures. The appropriate amount of warmth depends on the origin of the ancestral species; plants with ancestors from lowland areas near the equator (for example, *Arachnis* species) require more uniform conditions of warmth than do plants whose ancestral species came from higher latitudes or elevations. They should have abundant moisture—more when it is sunny and hot; less when it is cool and overcast—but they do not grow well if they are subjected to either prolonged stagnant wetness or persistent aridity. They should have bright light, but not of such intensity that the temperature of any of the leaves rises much above the temperature of the surrounding air. In short, they grow best when given bright but filtered light, ideally throughout the day. Long days of 12 hours or more, and only moderate seasonal variation in the length of the day, are optimal; obviously, growers in temperate zones cannot duplicate this last condition and are handicapped thereby. When day length is much less than 12 hours (the day length at the equator), the rate of growth will be slowed; longer day lengths, however, do no harm. Steady movement of air around the leaves and roots is highly beneficial, but the chilling effects of rapid evaporation at low temperatures or the desiccating effects of constant currents of hot, arid air are deadly. And, finally, the right climatic conditions must be combined with good nutrition and irrigation water of good quality.

Steady climatic and cultural conditions are best. Rapidly changing conditions lead to leaf drop and other manifestations of stress, because the plants cannot adapt to abrupt changes quickly enough. This sensitivity poses a particular challenge in geographic locations where the seasonal and daily changes in weather are great.

How can you tell whether your plants are being provided with a good environment? There are two main indicators: roots and leaves. The roots of a healthy, thriving plant are abundant and preferably branched. They are plump, with bright green tips (purple or reddish brown in some cases) varying from 0.6 to 2.5 cm (0.25 to 1 in) in length, depending on the time of year. At the onset of winter, the roots of many plants become dormant and have very short green tips or none at all. The second main indicator of a good environment is the degree of turgidity of the leaves. They should feel firm to the touch—not limp or desiccated.

When the environment is really bad, a number of plants will display signs of marked stress and decline, such as dead or dying roots and loss of lower leaves. The more difficult problem is to determine whether your plants are growing to their potential. Conditions that are merely mediocre, or a one-time shock such as a temperature drop to near freezing for a few hours, may have a retarding effect but cause no immediately visible symptoms of distress.

Read the awards descriptions in the *Awards Quarterly* of the American Orchid Society, and compare your plants with the awarded plants. Do your best ones have as many flowers? Are they as large? Are they as flat and well proportioned? The awarded plants may be superior genetically, but they also must be well grown in order to express any genetic superiority. If the flowers of your best plants consistently are smaller, fewer in number, and poorer in shape and color, they probably are not being grown as well as they might. The explanation is more likely to be found in the environmental conditions of light, temperature, moisture, and air circulation than in supplementary cultural practices such as fertilizing.

The climate standards and ranges recommended in this chapter are guided by the close to ideal conditions found in Chiang Mai (18°47'N), northwestern Thailand, where vandas and ascocendas are grown to perfection. The city of Chiang Mai has an altitude of 314 m (1030 ft). The tables in this chapter provide meteorological data for that city and for some other locations, for purposes of comparison. One of the locations listed may be reasonably similar to your own in latitude, day length, and climate, and may help you to understand your own seasonal problems better.

The guidelines provided in this chapter must be recognized as being only approximate, and not completely inflexible. Vandaceous orchids are grown reasonably well in many different geographical locations where the climate is not ideal. Most of the handicaps can be minimized or overcome if they are well understood. The practices recommended in this chapter will help you to emulate climatic conditions in Chiang Mai to the maximum extent feasible, and in most cases, that will be enough to assure success in growing vigorous plants with good flowers.

# LIGHT

The single most important environmental factor is solar radiation, the natural source of light and heat. Light is essential for photosynthesis, and photosynthesis is essential for plant growth. Photosynthesis is the process by which plants convert solar energy to food energy. The amount of food energy produced by photosynthesis is in direct proportion to the amount of light a plant receives, so long as the intensity of the light is not so great as to create damaging levels of heat.

For horticulturists, solar radiation has two aspects: (1) its *intensity* at any given moment; and (2) its total *amount* during each day. For vigorous growth and frequent flowering of your vandas and ascocendas, you should try to provide an amount of sunlight as close to the ideal as possible throughout the year; however, at no time should the intensity of light be too strong.

These goals are not easy to achieve. In the winter months, the days in the temperate zones are too short, and during the rest of the year, most growers tend to provide less light than is optimal and available (for example, by applying too much shading, or by applying shading too early and leaving it in place too late, or by not keeping their greenhouse glazing clean). On the other hand, every grower encounters brief periods of unseasonally strong light and high temperature that, despite the use of an amount of shading normal for the season, would overheat and scorch the foliage of the most exposed plants unless they are misted frequently during such episodes.

The greater the number of days and weeks in which lighting conditions in a temperate zone greenhouse are close to optimal during spring, summer, and early fall, the less harmful will be the unavoidable extended periods of insufficient sunlight and warmth in the late fall and winter months when days are short and plant growth is minimal. Plants that enter their winter dormant period in robust condition are capable of emerging from it with a burst of growth in the spring; those that do not enter that period in such condition will respond much more slowly to the surge in light and warmth when winter ends. To produce top quality plants and flowers, shading must be applied with considerable skill during the months in which it is needed, and it should be diminished or removed as soon as it is not needed. Timing is everything.

The maximum total potential amount of light a grower can provide for his or her plants during any season of the year depends primarily on the maximum intensity of sunlight in the region at noontime, when solar radiation normally reaches its peak, and on the number of hours between sunrise and sunset (i.e., day length) for the pertinent time of the year. Both of these factors are highly seasonal and are heavily influenced by the latitude of the grower's location.

In temperate climates, the increase in the total daily amount of light brought about by the expanding number of hours between sunrise and sunset in the first half of the year does no harm to vandas and ascocendas. Rather, greater day length is beneficial, so long as it is not accompanied by light and heat so intense as to desiccate plants. Moreover, the daily changes in day length are predictable.

The occurrence of days of excessively strong light follows an unpredictable and erratic pattern that depends on daily local weather conditions as well as on the seasonal trend line. Spring and fall are difficult seasons for temperate zone orchid growers, and test their ingenuity and perseverance. The trend line for the season tells us that there will be a surge in average daily solar radiation as spring approaches, and a pronounced decline early in the autumn, but growers have to deal with sunlight on a day-to-day basis rather than on the monthly daily average; some springs are marked by many days of inadequate light and some autumns are marked by days of unusual brightness and warmth. By summertime, sunlight almost everywhere is too strong from midmorning until late afternoon, except for intervals of rain, fog, or heavy clouds. High summer temperatures aggravate the situation. In the wintertime in temperate zones, there are few days of light so intense that shading is needed for vandas and ascocendas, and low outside temperatures prevent excessive buildup of heat in the greenhouse.

Within limits, growers can affect what will be the normal maximum intensity of sunlight in their greenhouse, up or down, by the siting* and design of the structure, by the placement of their plants in it, and by the choice and cleanliness of the glazing material. They can, however, as a practical matter, do nothing about day length, and in temperate zones, day length is too short for optimum total available light in the late fall and winter months, even though, on some days, the intensity of sunlight at noonday is satisfactory.

Growers sometimes are misled by the brightness of the light outdoors and underestimate the loss in transmission through the greenhouse glazing. Actual working values of light transmission through greenhouse glazing typically range no higher than 55% to 65% of outdoor readings, regardless of the type of glazing used; average transmission is only about 68% on the best new, large-pane commercial glasshouses (Firth 1992). The reasons for the reduced transmission are algae, dirt, and residues of chemicals on the covering, as well as the angle of the surface of the covering material in relation to the sun, and structural blockage.

### Using a Light Meter

In order to be able to control the maximum brightness of light in your greenhouse with any precision, you must have a way of measuring brightness. The intensity or the brightness of sunlight at any given moment can be measured moderately well with some kinds of photographic light meters. Although they are capable of mea-

*Because of the low angle of the sun's rays at locations above 38°N, solar radiation entering a f standing greenhouse will be maximized, and deflection by the glazing minimized, if the ridge ru east–west; below that latitude, the rays are high enough to make a north–south orientation bes

suring only the visible wavelengths of solar radiation (i.e., the wavelengths to which the human retina responds), those are the wavelengths involved in the process of photosynthesis. The visible portion of sunshine comprises about half of total sunlight; nearly another half is comprised of infrared rays, and the very small remainder consists of ultraviolet rays. These latter two play hardly any role in the rate of photosynthesis.

Within the visible band, all wavelengths are not equally effective in prompting photosynthesis. The orange, short-red region of wavelengths produces the maximum rate of photosynthesis, with a secondary maximum rate in the blue region. The green and yellow wavelengths produce a lower but still appreciable rate of photosynthesis. Light meters cannot measure the intensity of the individual sections of the visible portion of sunshine—only the total, but that is sufficient information for greenhouse and outdoor growers. For those who grow under lights, however, the wavelength distribution of their artificial illumination is important, because different bulbs have different patterns.

I recommend light meters that measure reflected light along with direct light, and that have a scale with a range from 0 to at least 10,000 fc. The scale should be expressed in footcandles and should not require a cumbersome conversion from some other unit of measurement. Use the meter frequently, and in many different areas of the greenhouse. You probably will be amazed at the variations, and you must take them into account when deciding on the amount and timing of shading, and on the placement of your plants, misting nozzles, and fans.

## Optimal Light

The specific goals to pursue and the rules to follow are quite straightforward.

1. Make any changes you can that will assure maximum intensity of close to 6000 footcandles (fc) in the brightest part of your growing structure (usually this will be on the south side, near the tops of the highest plants) for as many hours of the day and for as many days of the year as possible;
2. provide at least 2000 fc of light at midday to each plant, if you can;
3. apply shading whenever and wherever needed to prevent readings of higher than 6000 fc, but try to do it in a way that will minimize the lowering of light intensity elsewhere.

Light changes from hour to hour and from day to day, and the application of shading is a time-consuming chore. Obviously, compromises have to be made. Make them in the direction of more light rather than less light, but only up to the point where scorching and desiccation of leaves would begin to occur and could not be remedied by rearrangement of the exposed plants, or by timely misting. Whenever the danger point is reached, attend to the problem immediately. If not, harm may be done in a matter of a few hours.

Will observance of the three rules provide enough total light for optimal growth of your plants throughout the year? In most zones above 40° latitude, probably not, but skillful management of the controllable variables should permit very satisfactory growing and blooming of strap-leaf vandas and ascocendas almost everywhere except during the late fall and winter months. Because of reduced photosynthesis, these months are a resting period for your plants. Be sure to take this into account in your fertilizing and watering program.

In temperate zones, in order to keep light intensity for all vandas and ascocendas in your greenhouse within a range of from 2000 fc to 6000 fc during midday hours on sunny days, you would have to apply shading materials in a selective and graduated fashion as solar radiation increases seasonally, and you would have to reduce shading selectively and gradually in the latter part of the year as solar radiation diminishes. In practice, growers in temperate zones have to settle for something less than the ideal, and each hobby grower has to work out his or her methods for coping with the prevailing circumstances to the extent feasible, but a well-conceived schedule for applying and removing shading is a vital part of any sensible program.

## Solar Intensity and Latitude

Although the midday intensity of sunlight varies from day to day, there is an overriding seasonal pattern in the daily average intensity and in the total quantity of solar radiation. The seasonal pattern is related to the latitude of the location. In designing a program for dealing with excessive brightness, an understanding of the normal seasonal pattern can provide a framework for the practical decisions that have to be made.

Under conditions of cloudless sky and constant transparency, on any given day, the maximum number of hours of sunshine for any unshaded outdoor site, and both the peak and the average brightness of solar radiation on that site, are determined by the distance of the site from the sun and by the angle of the sun in relation to the site. There are no places on earth, however, where the sky is cloudless and of constant transparency. Potential brightness always is modified by local atmospheric conditions—weather patterns, water vapor, dust, suspended particles of pollutants, and reflections from snow, water, or other reflective surfaces. Geographical latitude, therefore, provides a good but not perfect indication of what lighting conditions a grower should expect at different times of the year in his or her area.

At the vernal and autumnal equinoxes (about March 21 and September 23, reversed in the Southern Hemisphere), the noonday sun is directly overhead at the equator, and solar intensity is exceedingly strong there. At the summer solstice (about June 22), the noonday sun is directly overhead at the Tropic of Cancer

(23°30' north of the equator); at the winter solstice (about December 22), it is directly overhead at the Tropic of Capricorn (23°30' south of the equator). These two latitudes are the outer reaches of the band of latitudes within which the noonday sun on one day of the year is directly overhead, and the sun will be nearly overhead throughout the year. For the year in its entirety, solar radiation is greatest at the equator, which is midway on the band.

The farther a location is from the Tropics of Cancer (in the Northern Hemisphere) and Capricorn (in the Southern Hemisphere), the greater the seasonality in the intensity of solar radiation, due to greater variation in distance from the sun and in the angle of the sun's rays. That greater seasonality makes it more difficult for orchid growers in temperate regions to provide the optimal seasonal amount of shading, because that amount changes at a significant rate from one month to the next. In the tropics, on the other hand, light intensity on sunny days, and during sunny hours in the rainy season, is substantially uniform throughout the year.

### Light Intensity in Chiang Mai

Chiang Mai, where lighting conditions are nearly ideal, lies well within the band established by the Tropics of Cancer and Capricorn; consequently, on clear days, both maximum and average solar intensity there are very high. Thai nurseries provide 50% shading throughout the year with permanently installed wood lath running in a north-south direction and placed at least 1 m (3 ft) above the tops of the mature plants (see Plate 99). I have taken light-meter readings at several Thai nurseries on sunny mornings and afternoons, between 10:30 A.M. and 3:30 P.M. In none of the samples did footcandle measurements at the level of the most exposed plants exceed 6000 fc. A range of 3500–4500 fc at mid-level of the plants was typical of nurseries in Chiang Mai. Some plants in all nurseries were receiving 2000–2500 fc at the time of the reading, and looked to be in fine condition; they might possibly receive more light at some other times of the day, when posts or other obstructions were not partially blocking the sun's rays. The goals I have set for optimal intensity of light are based largely on these findings; to the extent I have been able to achieve them, they have yielded good results for me, too.

In midsummer, Chiang Mai has considerable cloudiness and rain, which substantially reduce the average intensity and total quantity of light. During those months, local orchid growers sometimes are plagued by fungi and bacteria—even Chiang Mai is not *Vanda* heaven on earth.

### Day Length and Latitude

Day length is a major factor in determining the total amount of solar radiation a grower's plants will receive. That amount, rather than just the peaks in brightness, is what governs the amount of photosynthesis that will occur. The greater the number of hours of sunshine each day, the lower the tolerable *average* level of brightness compatible with good growth.

For the calendar year as a whole, average day length is the same everywhere—approximately 12 hours. At the equator, the number of hours between sunrise and sunset is 12 hours throughout the year. Above and below the equator, the number of daylight hours is seasonal; day length is longest at the time of the summer solstice and shortest on the date of the winter solstice. At the time of the vernal and autumnal equinoxes, when the sun crosses the equator, day length is 12 hours everywhere. The greater the distance from the equator, the greater the seasonality in day length.

Table 8-1 shows the relationship between latitude and day length very clearly. To relate the latitudes to illustrative locations in the United States, refer to Table 8-2. Note that at 48°N, roughly the latitude of Seattle, Washington (or Stuttgart, Germany), average day length in June is 16 hours, or 33% more than at the equator, and 3 hours, or 25%, more than in Chiang Mai, which has a latitude of 18°47'N. But in December, average day length at Seattle's (or Stuttgart's) latitude is only about 8½ hours, while it is 11 hours at the latitude of Chiang Mai. Obviously it is not possible to get as good growth in months with 8½-hour days as it is in months with 11-hour days.

Day lengths of at least 11 hours are ideal—that condition is met in the tropics where the wild species of vandaceous orchids are indigenous. Unfortunately, growers who reside above about 40° latitude have appreciably shorter days in the winter months than are optimal. Their problem is compounded by the fact that the av-

Table 8-1. Mid-month day length at certain northern latitudes (in number of hours and minutes) for 1987.

| LATITUDE | JAN. | FEB. | MAR. | APR. | MAY | JUN. | JUL. | AUG. | SEP. | OCT. | NOV. | DEC. |
|---|---|---|---|---|---|---|---|---|---|---|---|---|
| Equator (0°) | 12:08 | 12:07 | 12:06 | 12:06 | 12:07 | 12:07 | 12:08 | 12:07 | 12:07 | 12:07 | 12:07 | 12:07 |
| 10°N | 11:36 | 11:49 | 12:04 | 12:22 | 12:35 | 12:43 | 12:38 | 12:28 | 12:11 | 11:53 | 11:41 | 11:33 |
| 20°N | 11:02 | 11:29 | 12:00 | 12:36 | 13:03 | 13:20 | 13:12 | 12:51 | 12:16 | 11:40 | 11:13 | 10:56 |
| 30°N | 10:24 | 11:08 | 11:57 | 12:54 | 13:37 | 14:04 | 13:52 | 13:17 | 12:22 | 11:25 | 10:41 | 10:13 |
| 35°N | 10:02 | 10:55 | 11:55 | 13:05 | 13:58 | 14:27 | 14:16 | 13:31 | 12:26 | 11:16 | 10:23 | 9:49 |
| 40°N | 9:36 | 10:41 | 11:53 | 13:17 | 14:21 | 15:00 | 14:42 | 13:49 | 12:30 | 11:06 | 10:02 | 9:21 |
| 42°N | 9:26 | 10:35 | 11:53 | 13:23 | 14:31 | 15:13 | 14:54 | 13:57 | 12:32 | 11:02 | 9:52 | 9:08 |
| 44°N | 9:14 | 10:29 | 11:51 | 13:28 | 14:42 | 15:28 | 15:07 | 14:05 | 12:34 | 10:58 | 9:42 | 8:55 |
| 46°N | 9:00 | 10:21 | 11:51 | 13:36 | 14:54 | 15:43 | 15:21 | 14:12 | 12:36 | 10:53 | 9:32 | 8:40 |
| 48°N | 8:46 | 10:14 | 11:50 | 13:41 | 15:07 | 16:01 | 15:37 | 14:24 | 12:38 | 10:48 | 9:20 | 8:32 |

**Source:** Calculated from data in *The Astronomical Almanac for the Year 1987*, National Almanac Office, U.S. Naval Observatory, U.S. Government Printing Office: A14–A21.

erage intensity of solar radiation also is lower during the winter months. The result is that the total quantity of light is likely to be deficient, even without any shading and with clean, clear glazing. Growers in temperate zones can do nothing to alter these winter shortfalls in day length, which is why it is essential to try to maintain adequate brightness in the winter months (e.g., by avoiding use of darkening insulating coverings) and to maximize the growth of their plants during the months when the day length in their locale is equal to or greater than in the tropics.

## Total Quantity of Light

What really controls photosynthesis and plant growth is the *total quantity* of solar radiation a plant receives during the day. This amount is the product of day length and average solar intensity. Greater day length, within limits, can compensate for lower average intensity, but nothing (other than artificial lighting) can compensate for inadequate daily total solar radiation.

Measurement of total solar radiation during any specific period requires special calorimetric instruments not available to hobbyists. These instruments measure the quantity of solar energy by its heat-producing effects. One unit of measurement is a langley (British thermal units and joules are alternative units). A langley is defined as a unit of solar radiation equivalent to one gram calorie per square centimeter of radiated surface. A gram calorie, or simply calorie, is the amount of heat required at sea level to raise the temperature of one gram of water one degree Centigrade. For example, if the amount of solar radiation experienced by a square centimeter of leaf surface during the span of a given day was 300 langleys, that amount of solar energy would have been sufficient to raise the temperature of 10 grams of water 30°C (54°F).

The latitude of a given site determines the seasonality of its total quantity of solar radiation at different times of the year, and of the corresponding average hourly concentration of the energy. Table 8-2 shows the average daily readings of total solar radiation for each month at a number of urban locations in the United States, in descending order of degrees of latitude. Chiang Mai, our chosen standard for comparison, also is presented. The month of December is given first, because that is the month when the total quantity of solar radiation is lowest. Table 8-3 converts the daily data of Table 8-2 into daylight hourly averages for each month (i.e., total daily quantity divided by day length), but for a smaller number of locations which are spread roughly 10° of latitude apart. These hourly data are a measure of the monthly variations in the average hourly concentration, or total intensity of sunlight (i.e., not just of the visible portion of the spectrum, which is all that a light meter registers). This information, by providing an hourly average, can supplement the footcandle measurements a grower takes of the brightness of sunlight in his or her growing area, which, it must be remembered, are only momentary measurements.

Table 8-2. Average daily global solar radiation in langleys at selected locations—1952–75.

| LATITUDE | LOCATION | YEARLY AVG. | DEC. | JAN. | FEB. | MAR. | APR. | MAY | JUN. | JUL. | AUG. | SEP. | OCT. | NOV. |
|---|---|---|---|---|---|---|---|---|---|---|---|---|---|---|
| 47°27'N | Seattle/Tacoma | 286 | 57 | 71 | 134 | 230 | 351 | 465 | 489 | 610 | 438 | 311 | 178 | 92 |
| 42°22'N | Boston | 300 | 109 | 129 | 193 | 276 | 360 | 440 | 493 | 475 | 403 | 342 | 241 | 136 |
| 41°47'N | Chicago | 330 | 109 | 138 | 206 | 300 | 396 | 485 | 544 | 527 | 466 | 367 | 263 | 153 |
| 41°24'N | Cleveland | 296 | 86 | 105 | 163 | 250 | 366 | 456 | 500 | 496 | 429 | 336 | 235 | 126 |
| 40°46'N | New York (LGA) | 318 | 124 | 149 | 216 | 303 | 395 | 459 | 489 | 484 | 430 | 347 | 258 | 161 |
| 38°45'N | St. Louis | 360 | 144 | 170 | 240 | 327 | 424 | 508 | 568 | 556 | 493 | 396 | 298 | 195 |
| 37°44'N | Oakland | 416 | 176 | 192 | 276 | 395 | 521 | 600 | 637 | 630 | 557 | 461 | 329 | 223 |
| 36°12'N | Tulsa | 373 | 179 | 199 | 265 | 354 | 435 | 494 | 548 | 551 | 506 | 400 | 316 | 224 |
| 33°56'N | Los Angeles | 432 | 230 | 251 | 329 | 439 | 529 | 559 | 575 | 626 | 564 | 456 | 357 | 272 |
| 33°39'N | Atlanta | 365 | 183 | 195 | 263 | 354 | 457 | 503 | 519 | 492 | 463 | 386 | 326 | 240 |
| 29°59'N | Houston | 367 | 198 | 210 | 281 | 352 | 413 | 481 | 515 | 496 | 457 | 399 | 346 | 251 |
| 25°48'N | Miami | 400 | 276 | 287 | 356 | 435 | 504 | 500 | 463 | 478 | 442 | 395 | 353 | 303 |
| 21°20'N | Honolulu | 445 | 307 | 320 | 379 | 440 | 487 | 529 | 544 | 543 | 533 | 491 | 418 | 343 |
| 18°26'N | San Juan, P.R. | 445 | 335 | 360 | 417 | 485 | 513 | 492 | 493 | 508 | 499 | 454 | 411 | 371 |
| 18°47'N | Chiang Mai | 439 | 417 | 393 | 439 | 423 | 494 | 507 | 445 | 366 | 457 | 448 | 448 | 431 |

**Explanatory Notes:** Total Global Solar Radiation reported in this table is the average daily sum of direct, diffuse, and ground-reflected solar radiation on a horizontal surface. A langley is a unit of solar radiation equivalent to one gram calorie per square centimeter of irradiated surface.
**Source:** *Insolation Data Manual and Direct Normal Solar Radiation Data Manual,* published by the Solar Energy Research Institute, July 1990, Golden, Colorado. For Chiang Mai, Thailand, the data came from Löf, George O. G., et al., *World Distribution of Solar Energy, Report No. 21,* 1966, Univ. of Wisconsin Solar Energy Laboratory: 47.

Table 8-3. Average daytime hourly global radiation in langleys at selected locations— 1952–75

| | DEC. | JAN. | FEB. | MAR. | APR. | MAY | JUN. | JUL. | AUG. | SEP. | OCT. | NOV. |
|---|---|---|---|---|---|---|---|---|---|---|---|---|
| Seattle 47°27'N | 7 | 8 | 13 | 20 | 26 | 31 | 31 | 40 | 30 | 25 | 16 | 10 |
| New York 40°46'N | 13 | 16 | 20 | 25 | 30 | 32 | 33 | 33 | 31 | 28 | 23 | 16 |
| Houston 29°59'N | 19 | 20 | 25 | 30 | 32 | 35 | 37 | 36 | 34 | 32 | 30 | 24 |
| Honolulu 21°20'N | 28 | 29 | 33 | 37 | 39 | 40 | 41 | 41 | 40 | 36 | 31 | 28 |
| Chiang Mai 18°47'N | 38 | 35 | 38 | 35 | 39 | 39 | 34 | 28 | 36 | 37 | 38 | 38 |
| (with 50% shading) | (19) | (18) | (19) | (18) | (20) | (20) | (17) | (14) | (18) | (19) | (19) | (19) |

**Source:** Calculated from data in Tables 8-1 and 8-2.

The latitude of your location is a factor that can help guide you in designing a program for applying shading and other environmental controls. The negative effects of local atmospheric conditions, however, can be substantial. Those conditions have their own seasonality, and they can materially modify the amounts of light that would be expected from the influence of latitude alone. Some practices that produce good results for a grower at the same latitude elsewhere may not be best for you in your region.

For example, in Seattle, only 28% of potential solar radiation reaches the ground in January, but 63% does in July (this accounts for the high hourly average for July shown in Table 8-3). In Miami, the corresponding monthly average is approximately 50% throughout the year. In New York City, the monthly percentages range from 34% in December to 47% in July. In San Juan, Puerto Rico, the range is from 52% in June, September, and October to 57% in March. The merit of the monthly data on total solar radiation in Tables 8-2 and 8-3 is that they show the actual average solar radiation, and therefore record the combined effect of latitude, season, and prevailing atmospheric conditions at the location during a typical stated month.

Notice in Table 8-2 that despite the influence of local factors, the higher the latitude the greater the seasonality and the smaller the daily average total amount of solar radiation for the year as a whole. For example, Cleveland, with a latitude of 41°24'N, on the average gets 25% less total solar radiation each year than does Miami, which has a latitude of 25°48'N—an average daily amount of 296 langleys vs. 400 langleys for Miami. Chiang Mai (18°47'N) receives a daily average of 439 langleys, with only moderate differences in the daily averages for most months of the year, whereas the range for Cleveland is very wide—from an average daily high of 500 langleys in June to a low of 86 in December. Seattle shows an even greater stretch.

Table 8-2 also illustrates that, along with latitude, such elements as cloud cover, air pollution, and proximity to large reflecting surfaces of water play a role in the amount of global solar radiation a site receives. For example, Oakland, California, located on the San Francisco Bay, is more than 4° farther north than Atlanta, Georgia, which is inland, yet Oakland's average daily global solar radiation is 416 langleys in comparison with Atlanta's 365. One factor in the Oakland measurement is the light reflected from the Bay.

In temperate zones, growers of vandas and ascocendas have no choice but to grow their plants under solar conditions quite different from those prevailing in the habitats of the ancestral species. For example, the latitude of the province of Davao in the Philippines, the habitat of *Vanda sanderiana*, is 7°N; *V. coerulea*, the northernmost species widely found in the ancestry of current *Vanda* and *Asco-*

*cenda* hybrids, comes from a latitude of approximately 19°N, which is about the same as that of Puerto Rico and not much lower that that of Honolulu (21°20'N). At these low latitudes, swings in solar radiation, both relatively and absolutely, are much smaller than at higher latitudes.

In northern latitudes above, say, 37°, the quantity of solar radiation more than doubles from December's lows during the first three months of the year, and a corresponding decline occurs in the last three months of the year. The average hourly concentration of solar radiation shows a similar pattern (see Table 8-3). During the months of early spring and late fall in the upper latitudes, lighting conditions must be monitored with a light meter from week to week; otherwise, suddenly excessive light early in the year will cause rapid desiccation of leaves accustomed to winter's much lower levels of intensity, and, in the early fall, unless alleviated by removal of shading, the sharp drop in intensity will produce vegetative softness that is vulnerable to bacterial and fungal disease.

In the Southern Hemisphere, comparable seasonal conditions and consequences apply, but the dates are six months different.

## Shading Materials

Most hobby growers are limited to a few rather inflexible ways of providing shading. Perhaps the most common is to place shade cloth, a woven plastic netting, over the top of the greenhouse or outdoor enclosure. Another very common method is to paint the exterior surface of the glazing with a removable whitewash made for greenhouses. Both materials are a nuisance to apply and remove.

Shade cloth is sold by many horticultural suppliers. It is available in various stated percentages of shading provided. My experience is that the degree of shading is understated in some cases, and I am puzzled about how manufacturers do their testing. For that reason, I recommend buying a sample of a square meter or yard and testing it before buying the total quantity needed. For a modest charge, suppliers usually will bind the edges to prevent unraveling and will insert grommets to facilitate tying down the netting, which must be done to keep it in place.

The usual colors of shade cloth are black or green. The filaments of these absorb and transmit heat—a disadvantage because some of the heat is transmitted into the greenhouse. Some suppliers carry a knitted white shade cloth. This type has the advantage of reflecting some light, which reduces the temperature inside the greenhouse significantly; because of that, a lower shade-rating will serve. The knitted plastic fabric also has the advantage of not having to be bound on the edges to prevent unraveling. Inexpensive plastic "grippers" are sold for use in tying down the fabric; by squeezing the two halves together, a loop is formed through which a cord can be passed.

A system of rollers and pulleys to raise and lower exterior-mounted shade cloth would be feasible but rarely is seen. Some commercial greenhouses have electronically controlled interior curtains that slide above the growing area whenever shading is called for and pull back when it is not needed. Such systems, combined with high-mounted exhaust fans, are ideal, but beyond the reach of most hobby growers. Besides, few hobby greenhouses have the height to permit an installation of this type.

Consider applying shading material to only certain sections of your greenhouse, such as the south side, and applying different (or no) shading to some other sections of the structure. In very few locations will shading of more than 40% ever be needed for greenhouse application; take this into account when buying shading material. Greater flexibility can be achieved by using narrower strips of shading fabric (if that is your chosen material) rather than a wide, single piece that would cover the entire greenhouse. Try to find a way to elevate shade fabric above the glazing, so that cooling air currents can pass between the underside of the fabric and the exterior surface of the glazing. This will reduce the amount of heat buildup within the greenhouse on hot days; with less risk of overheating, a lower degree of shading may be used, thereby providing more light for your plants. Many growers make excessive use of shading in order to control summer heat, but in the process they reduce the average level of brightness below what is ideal; such growers would be better served by making greater use of misting to control heat. For plants grown outdoors in the summertime, a shadier material will be needed than for greenhouse growing, because of the absence of glazing. Light-meter readings can be used to determine the extent of the difference.

A simple way to achieve a cooling effect is to arrange lengths of lightweight, flexible plastic drain tubes of about 10 cm (4 in) diameter under the shade cloth. The tubing is sold by most building supply outlets and is very inexpensive; its normal use is for carrying away rainwater from gutter downspouts. Arrange the tubes parallel to one another, about 75–90 cm (30–36 in) apart, either straddling the ridge (on free-standing greenhouses) or with one end fastened to the peak (on lean-to structures). Attach the shade cloth to the greenhouse in the normal fashion. The air space between the exterior surface of the glazing and the underside of the shade cloth creates a chimneylike effect that produces a draft between the two surfaces even though the netting sags somewhat in between the lengths of tubing (unless a way is devised to stretch it taut). This ingenious and simple solution was invented by Arthur Moore of Marysville, Tennessee.

An alternative to shade cloth is whitewash made for greenhouse use. It can be applied with a brush, roller, or sprayer. Some growers find the whitewash not too burdensome to apply in the spring and to scrub off in the fall. A succession of light coatings can be applied as the season progresses. That provides some control, but it is not very precise. Rain gradually reduces the opacity once new coats cease to be added; that is an advantage as fall approaches.

## Optimizing Light for Each Plant

For an individual plant, the only lighting conditions that matter are those that affect the climate in the special niche the plant occupies. At any given moment during the day, it is inevitable that there will be a variety of lighting conditions in any greenhouse full of plants, and the patterns will change during the day, but that should not prevent a reasonable amount of balancing for each plant over the day as a whole. Make light-meter tests at frequent intervals until you learn what the patterns are, and study them for what they reveal. Rearrange plants periodically as conditions change. See the sections on symptoms of excessive light and of insufficient light following this section.

In arranging plants, give the plants with the greatest potential the best locations. Place the strongest ones in the brightest locations (usually a southern exposure), and the youngest and weakest plants in the shadiest ones (usually an eastern or northern exposure). Never give a plant with poor roots much light; the plant simply cannot absorb as much moisture as it loses through transpiration, especially in warm weather or during the height of the heating season. Above all, do not overcrowd.

When the flower buds of a *Vanda* or *Ascocenda* begin to swell and open, the plant becomes more easily stressed by intense light and high temperatures—those conditions can cause dehydration even when the plant seems to have an adequate mass of roots. Flower quality suffers markedly as soon as any dehydration begins to occur: petals and sepals become badly recurved, flower size is reduced, and flower colors are less saturated and bright. It is highly advisable, therefore, especially in the summertime, to move exposed plants in bud to conditions of more subdued light and less pronounced rises (and falls) in temperature.

Overcrowding—a common condition—makes light intensities so different from one plant to another that no light-meter reading is representative for any considerable proportion of the total plants. It is impossible to determine the right amount of greenhouse shading in such a situation, because there is no single appropriate value. Reducing or delaying the application of shading materials results in too much light for plants on the south side of the greenhouse and for the plants near the peak. Trying to remedy that problem by frequent misting may help the more-exposed plants but is likely to cause rot and other disease among the most crowded plants.

If, despite all advice to the contrary, you have great irregularity of lighting conditions because you overcrowd your greenhouse, then do not hang many plants up high where they will block light to a far larger number of crowded plants behind and below them. Place the largest plants on the north side of the greenhouse and hang them high; there they will not shade other plants. Greater air movement is needed in overcrowded conditions than when each plant has ample space around it. Remember this when deciding the number, size, and placement of fans.

## Symptoms of Excessive Light

Individual plants receiving too much light (in conjunction with too little water and air movement) usually show desiccation or other signs of stress. Leaves are not hard and turgid; they may show blanched spots, indicative of sunburn, typically at a point of inflection around the midsection of a leaf (because of the angle of that part of the leaf's surface in relation to the sun's rays). The color of the leaves is likely to be yellowish green, instead of the brighter green of Granny Smith apples. The plants also have, or will develop, shorter and narrower leaves than normal, or the two longitudinal halves of the leaves will fold together, in an effort to reduce the amount of leaf surface exposed to intense light. Relocating the overly exposed plant usually will suffice to prevent further damage. For care of damaged plants, see the section on treatment of badly stressed plants in Chapter 9.

## Symptoms of Insufficient Light

Plant growth is retarded by too little light as well as by too much. Sometimes the problem is a general one, caused by the weather or by the use of too much shading. Often it is caused by overcrowding; then its impact on individual plants can vary enormously and cannot be resolved simply by measures to increase the overall amount of light entering the greenhouse.

Plants that chronically receive too little light are susceptible to fungi and bacteria; they bloom less frequently, and the flowers are of poorer quality—smaller, fewer in number, and with weaker color and substance. The symptoms of too little light are leaves that are very dark green and noticeably longer and wider than those on plants of comparable ancestry located in a brighter situation. Pay particular attention to the size of the newest mature leaves, because they are most representative of recent and current conditions. The reason the leaves are larger is that an under-illuminated plant develops more receptor surface in order to capture the limited available amount of light—exactly the opposite of what a plant receiving too much light does.

The solution is to remove obstacles to the passage of solar radiation to poorly illuminated plants, or to shift the plants to a brighter location. Make any such changes gradually, over several weeks. A sudden change from subdued light to bright light may stress the plant and set it back, or even plunge it into a long-term decline.

## TEMPERATURE

Ideal ambient temperatures vary among vandaceous orchid genera and species, but the differences are small enough to be accommodated by careful selection of microclimates in most greenhouses. Some species, for example *Vanda dearei* and *Arachnis hookeriana,* will flower under conditions of uniform warmth. Some, such as *V. roeblingiana, Ascocentrum curvifolium,* the *V. cristata* complex, *V. coerulea* and *V. coerulescens,* require cooler nighttime temperatures than others, for example, *V. sanderiana.* The temperature optimums for the species ancestors are transmitted, in varying degrees, to their hybrid descendants.

Those vandaceous orchids that grow best under uniformly elevated temperatures, along with high humidity, thrive in the climate of Singapore. Examples are the *Arachnis* species and hybrids, *Papilionanthe* species and hybrids, and *Vanda dearei.* These types are not easy to grow and flower well in the temperate zones (north of the Tropic of Cancer and south of the Tropic of Capricorn). Most standard strap-leaf *Vanda* hybrids and ascocendas, however, perform better in the somewhat wider range of temperatures found in the Bangkok area. Even in Bangkok the daily and seasonal movements are not sufficiently pronounced to be ideal for *V. coerulea.* For that species, the climate of Chiang Mai in northwestern Thailand is ideal. There, *V. coerulea* and its hybrids can be grown to perfection, but so can strap-leaf vandas with heavy *V. sanderiana* ancestry. *Ascocenda* hybrids also grow well there, as do most of the other intergeneric hybrids with *Vanda* (e.g., *Christieara, Kagawara,* and *Vascostylis*).

The challenge to growers in temperate zones is to produce patterns of temperature in their greenhouses, especially in wintertime, that reasonably approximate those of Chiang Mai. That is an attainable goal; the acceptable range is about 16–21°C (60–70°F) for normal lows and 27–35°C (80–95°F) for normal highs. The midpoint of each range may be considered to be optimum. A high-low spread of at least 8°C (15°F) is highly desirable.

### Tropical Temperature Patterns

Comparing the temperature patterns of Singapore, Bangkok, and Chiang Mai offers some useful insights into the best temperature conditions for most vandaceous orchids. All three places provide sufficient sunshine and humidity. In each of them, air movement is mild but quite steady. There are, however, significant differences in the temperature patterns, as Table 8-4 demonstrates.

Table 8-4. Chiang Mai, Bangkok and Singapore temperatures:
average daily maximum, minimum, and spread.

| | CHIANG MAI | | | BANGKOK | | | SINGAPORE | | |
| | MAX. | MIN. | SPREAD | MAX. | MIN. | SPREAD | MAX. | MIN: | SPREAD |
|---|---|---|---|---|---|---|---|---|---|
| JAN. | 84°F | 56°F | 28°F | 89°F | 68°F | 21°F | 86°F | 73°F | 13°F |
| FEB. | 89 | 58 | 31 | 91 | 72 | 19 | 88 | 73 | 15 |
| MAR. | 94 | 63 | 31 | 93 | 75 | 18 | 88 | 75 | 13 |
| APR. | 97 | 71 | 26 | 95 | 77 | 18 | 88 | 75 | 13 |
| MAY | 94 | 73 | 21 | 93 | 77 | 16 | 89 | 75 | 14 |
| JUNE | 90 | 74 | 16 | 91 | 76 | 15 | 88 | 75 | 13 |
| JULY | 88 | 74 | 14 | 90 | 76 | 14 | 88 | 75 | 13 |
| AUG. | 88 | 74 | 14 | 90 | 76 | 14 | 87 | 75 | 12 |
| SEPT. | 88 | 73 | 15 | 89 | 76 | 13 | 87 | 75 | 12 |
| OCT. | 87 | 70 | 17 | 88 | 75 | 13 | 87 | 74 | 13 |
| NOV. | 86 | 66 | 30 | 87 | 72 | 15 | 87 | 74 | 13 |
| DEC. | 83 | 59 | 24 | 87 | 68 | 19 | 87 | 74 | 13 |
| YEAR | 89°F | 68°F | 21°F | 90°F | 74°F | 16°F | 87°F | 74°F | 13°F |

**Source:** *Tables of Temperature, Relative Humidity and Precipitation for the World*, Part V, *Asia* 1976.
Meteorological Office, Her Majesty's Stationery Office, London.

In Singapore, the spread between the average daily highs and lows for each month is stable and small throughout the year; it is practically the same as the yearly average high-low differential of 7°C (13°F). Those conditions are hard to approximate in temperate zones, even in the best-managed greenhouses. In Bangkok, the monthly spread for the year is wider (9°C [16°F]), and the month-to-month figures show greater movement; the deviations from the average fall within a range of plus-or-minus 1.7°C (3°F) from the annual average spread of 9°C (16°F), except for the month of January, when the spread is 12°C (21°F). The magnitude of the monthly high-low spread rises steadily from the end of October through January. Early October marks the end of the rainy season, which starts in May; November through April are the relatively dry months, and the nights are cooler until March.

Note that the average monthly daily low for the year is identical for Singapore and Bangkok. For most months, the wider high-low range for Bangkok reflects hotter daytime temperatures there than are characteristic of Singapore, rather than lower nighttime temperatures. It is not surprising, therefore, that many kinds of orchids grow about equally well in both places. While there is indeed greater seasonality in Bangkok's temperature pattern, the differences are moderate.

Much more seasonality in the monthly spreads is apparent in Chiang Mai's profile. The average daily lows for most months are appreciably below those of Bangkok or Singapore, and the high-low ranges, except for a couple of months at the peak of the rainy season, are markedly wider. They are wider not because the daytimes are cooler than in Bangkok—except for December and January, Chiang Mai gets as hot as Bangkok in the daytime—but rather because Chiang Mai is appreciably cooler at night during most months of the year. *Vanda coerulea, V. denisoniana,* and *Ascocentrum curvifolium* do better under the cooler nighttime conditions, and their hybrids with *V. sanderiana, V. dearei, V. luzonica,* and *V. tricolor* do at least as well in Chiang Mai as in Bangkok and Singapore in most cases—and often better.

During most of the year, greenhouse growers in temperate zones can easily duplicate Chiang Mai's minimum temperatures, and at the same time can achieve a satisfactory, though sometimes narrower, high-low range.

## Controlling Ambient Temperatures in Winter

In the winter, try to achieve an average daily low in the 16–21°C (60–70°F) range. Most growers already provide that for all their orchids except the so-called cool-growing types, such as masdevallias and odontoglossums. On some sunny days, solar radiation may raise the temperature inside many greenhouses to 27°C (80°F) or more, for at least a few hours, without any resetting of the thermostat. If that does not occur occasionally, the explanation may be that you are using an insulating cover that excessively reduces light intensity.

Adequately high daytime temperatures on sunny days in winter months can be assured by raising the setting of the thermostat during the middle hours of the day (say, to 27°C [80°F]), without raising fuel costs very much. The goal is to raise the temperature only 2 or 3°C (4 or 5°F) during the warmest hours of the day (above what it would be in the absence of such action). This is not essential, but the somewhat higher midday temperatures will promote more growth. If you are much concerned about the extra costs, I would recommend lowering the nighttime setting of the thermostat to 16 or 17°C (62°F), to compensate for the additional consumption of fuel during the middle hours of the day. On dark days, not enough sunlight is available to promote photosynthesis, so there will be little growth even if heaters boost the daytime highs; of course, thermostats controlling heaters should never be set lower in the daytime than at night.

In the winter, localized differences of temperature inside greenhouses may be pronounced, and they may cause damage before they are detected. This situation may exist not only in periods of extreme cold during which the greenhouse heaters are running almost constantly; it also occurs at other times of the year, whenever the outside temperature is very high. In both sets of circumstances, temperatures at various elevations and sections inside the greenhouse may show amazing departures from their customary ranges. For example, plants very close to the glazing can get very cold (or hot), regardless of the temperature in the middle of the greenhouse.

Do not be alarmed if your heating system occasionally cannot maintain the nighttime temperature at the normal minimum. Remember, "minimum temperature" means the lowest temperature reached anywhere where plants are positioned. That is not necessarily the same as the minimum setting of the thermostat. Place maximum-minimum thermometers at various levels and places throughout your greenhouse and observe them frequently. A few winter nights when the greenhouse temperature drops into the 10–15°C (50–59°F) region will not do any harm. In fact, it is not uncommon for *Vanda* growers in southern Florida to experience such temperatures in their greenhouses in the wintertime. But if the greenhouse temperature consistently falls below 16°C (60°F), or below 10°C (50°F) at any time, the system requires attention. Something also needs to be done if the temperature anywhere plants are positioned falls below these minimums—perhaps some plants need to be moved farther from the exterior walls, or fans need to be redirected.

In subtropical areas where plants are kept outdoors most of the year, the plants usually are tougher than those grown in a greenhouse year-round, and they are able to withstand somewhat longer periods of lower temperatures in winter—even very brief periods of temperatures as low as 7°C (45°F). They will not thrive, but they are unlikely to suffer permanent damage. Much depends on whether they enter the winter season with hard, tough leaves; that, in turn, depends on the kind of environment and culture they were given during the summer.

Many northern growers cover the exterior or interior surfaces of their greenhouse glazing with some kind of insulating material, such as polyethylene film or "bubble plastic," to reduce heat loss in the winter months. Their intent is very understandable, but often two undesirable consequences result. First, the covering may block too much light at a time when vandas and ascocendas need all the sunlight they can get. Manufacturers' declarations about the high transmission of light by their plastic products are potential figures under ideal conditions (e.g., absolutely clean surfaces) and are never achieved after the material has been in place for even a few weeks. Second, solid sheetings may be applied in such a thorough way that insufficient fresh air enters the greenhouse, and plant growth suffers. While fuel is expensive, so are orchids; it is false economy to save money on fuel if the growth of your plants suffers. Recently I decided to eschew all insulation in order to obtain more light. My plants responded well to the additional light, even though my heaters sometimes could not maintain the nighttime temperature lows as high as I wished (i.e., 16°C [61°F]) because of unusually cold winter weather.

### Placement and Use of Equipment in Winter

The best location for the heating thermostat generally is in the center of the greenhouse, at about eye level. Do not place a thermostat adjacent to an outside wall, especially a south wall; the temperatures in the interior not only will differ materially from the thermostat setting, but they also will be subject to wide swings, depending on how cold it is outside and, in daytime, on the amount of solar radiation. While this should be obvious to anyone, apparently it is not so to some of the workmen who erect greenhouses and install equipment; they should know better but often do not. I say this from experience!

In cold climates, where the heating system frequently operates full blast, the air can become desertlike in locations near the heater or its pipes or ducts, even though temperatures and humidity are normal in most other areas of the greenhouse. The situation is even worse when the day is very sunny and cold. The heater can cause desiccation astonishingly quickly (sometimes in less than an hour) in plants receiving the full effect of the withering hot air, unless counteracted by misters and strong fans that sweep the vapor through the dry areas. In cold climates, the misters or foggers generally should operate constantly during the height of the heating season, regardless of the frequency of drenchings. Aim to keep daytime relative humidity in about the 60–80% range, so long as the air temperature is 18°C (65°F) or higher, and vigorous air movement is provided.

During the winter, the placement of fans may have to be quite different than in the summer. Use high-velocity fans, together with misting nozzles, to quickly disperse concentrations of hot air. Ceiling-fan motors should be reversed (if the fans have a reverse switch) so that the movement of the blades pushes the somewhat warmer upper air downward. Ceiling fans that are not reversible are designed only for use in hot weather. There is much less of a differential in temperature between the upper and lower levels of a greenhouse in the wintertime, so eliminating a layer of superheated air at the peak is not a problem; the cold exterior surface of the glazing, along with related condensation and evaporation on the inner surfaces, takes care of most of that problem. Instead, the objective is to prevent the heating system from causing localized areas of excessive heat and desiccation. The extent of the latter problem depends on the kind of heating system, and on how it was installed. Some installations are much less of a problem than are others. Fortunately, the necessary equipment needed to improve conditions is inexpensive and easily assembled by anyone with handyman's skills. Fine-tuning, however, requires careful observation and patience.

To keep the area above a source of heat usable for plants, mount a strip or sheet of aluminum corrugated roofing 30 cm (12 in) or more above the source of the heat; then mount a fine misting nozzle at the end of a garden hose so that it sprays onto the hot metal surface below. Place a fan nearby to blow the mist in an appropriate direction. Take care to avoid saturating nearby plants with this setup. If that is a problem, use a short section of dripper hose instead of a misting nozzle to wet the corrugated metal. Corrugated concrete roofing panels also can be used above sources of heat. They will not transmit heat, and therefore need not be kept wet; however, they may be difficult to find, and they are very heavy.

### Controlling Ambient Temperatures in Summer

To keep ambient temperatures in greenhouses from rising too high in the summertime, growers need to make skillful use of thermostats and humidistats to regulate vents, fans, misting apparatus, and evaporative coolers. The alternative is to move the plants outdoors.

Let the greenhouse temperature rise to about 29°C (85°F) before any evaporative coolers come into play, and set the controls for roof vents and exhaust fans a few degrees below that. Be aware that setting the controls at these high levels will almost certainly permit the temperature in the top section of the greenhouse to rise significantly higher than 30°C (86°F) on hot days, despite use of devices to improve ventilation. (Only heavy shading and a very powerful misting or fogging system that directs the flow of moisture-laden air upwards can prevent pronounced temperature buildup near the peak.) If moisture is kept high, plants normally will not be damaged by such high temperatures; quite the contrary, they will thrive. But if they are not watered rather frequently, they soon will be harmed. The need for frequent waterings under conditions of high temperatures does not obviate the corollary need for a drying interval to prevent or minimize algae, rot, and fungi, all of which thrive under warm, moist conditions.

Place a number of maximum-minimum thermometers in various places and elevations throughout the greenhouse to help you locate the areas where temperatures rise the most, and note the magnitude of the differences from one location to another. Also, be on the lookout for desiccation or scorching of plants hung high.

Thermostats that control cooling and ventilating equipment should be mounted in what normally is one of the hotter locations in the greenhouse, but not in the very peak. Be sure to direct the misting nozzles away from the thermostat; otherwise, the mist will cool the thermostat almost instantly, and the "on" cycle will be too short. If an automatic misting system shuts your thermostat off too soon, or keeps your plants wet too long, you can connect a timer to the system to ensure that the misting never stays on for more than, say, 15 seconds during any desired period.

The best venting system for preventing excessively high interior temperatures is a combination of thermostatically controlled roof vents and low-level intake openings, preferably located all around the base of the structure. This combination produces a chimney effect that provides a cooling draft of air from the lowest level of the greenhouse up through the roof; it should be supplemented with adequate shading of the structure.

An alternative is the use of a powerful exhaust fan or fans. These generally are placed in the vertical wall at one end of the greenhouse and have shuttered intakes at the other end. Although not always feasible, it is best to have the intakes mounted low and the exhaust fans mounted high—that reduces temperature differentials at different elevations in the greenhouse.

Wet pad systems and evaporative coolers, installed at ground level, also help to lower temperatures. These use powerful fans or blowers to force air over wet surfaces; they cool by evaporation. They are most effective if the relative humidity of the outside air is low, because that raises the rate of evaporation, which is what produces the cooling. If outside humidity is high, there is little evaporation and hence little cooling (unless a constant flow of new cool water is provided, instead of recirculated water). These cooling systems have the added advantage of raising humidity in the greenhouse, if it is not already high.

Many types of equipment are available for regulating air movement and ventilation in the summertime, and consequently the patterns of maximum temperatures within a greenhouse. You can choose from horizontal fans, high-velocity and low-velocity fans, large volume and low volume fans, ceiling fans, rotating fans, exhaust fans and intake fans. Each has its use. Some can be effectively combined with misting nozzles to provide very effective cooling and humidity. What matters most is how well you orchestrate the pieces of equipment.

Arrange equipment to produce the patterns of temperature you desire, insofar as that is possible. That is why frequent monitoring of plants and equipment is so important. Maximum-minimum thermometers are invaluable for this purpose.

Few efforts produce as high a return as the careful selection and placement of equipment to regulate the patterns of temperature (and moisture) in different areas and elevations of a greenhouse, especially in the hottest months. Compromises are unavoidable. While a grower of only Sarcanthinae species and hybrids ideally would like to have low-velocity air of uniform temperature throughout the greenhouse, a grower with a more mixed collection usually will want controlled variations in some parts of the house. In any case, it is desirable to use high-velocity fans near sources of moisture in order to produce rapid dispersion of vapor throughout the house and thereby avoid both hot pockets and cold, wet ones. In the summer, ceiling fans with reversible motors should be set to pull cooler lower-level air upward, rather than to push warmer air downward.

## Safety Standards for Heating and Cooling Equipment

Buy only equipment that can withstand the hostile environment of greenhouses and lathhouses. Electric heaters, sometimes used as supplementary sources of heat, can be especially dangerous. Make sure that wiring and outlets meet the standards of the electrical code. Use plastic conduit for the wiring, and plastic outlet boxes instead of metal ones. Fans sold at seasonal clearance sales at local hardware or appliance stores may not long withstand greenhouse conditions. Worse still, they may blow a fuse and shut down other vital equipment when you are not there to correct the situation, and you may then lose your orchids to extreme heat (or cold). Worst of all is the possibility of a lethal shock. Wetness and electricity do not go well together. Protect the electrical equipment, outlets, and switches from any direct exposure to water. Do not touch them when they are wet. The life you save may be your own or that of a loved one. Many localities now require the installation of ground-fault interrupters. Consider installing them even if they are not required by local ordinances (but shield them from moisture, else they will constantly be triggered).

## Placing Plants Outdoors in Summer

Many growers in temperate zones hang their vandaceous plants outdoors in the summer, mainly in order to avoid the problems of heat buildup in their greenhouse. With adequate shade cloth or other shading material overhead (preferably mounted at least 1 m [3 ft] above the tops of the tallest plants), leaf-scorching can be avoided. Nature attends to the need for air circulation. Automatic or semi-automatic watering or misting systems easily can be improvised to control extremes of temperature, simply by using a garden hose or plastic tubing with appropriate nozzles attached or rotating lawn sprinklers. The watering systems can be combined with thermostatic controls to provide protection against excessive heat and also against frost. Where neither of these potential threats is a problem, battery-operated timers, featured in many garden catalogs, can be attached at the faucet to provide timed wetting and cooling, with numerous on-off cycles. Another advantage to outdoor growing is that the plants generally can be spaced farther apart than is possible in a greenhouse. That assures better distribution of light—a great advantage.

Moving plants outdoors in the summertime has a few disadvantages. Carrying them outside and later bringing them back inside, sometimes with very short notice if the weather changes abruptly, can be burdensome. Insect attacks are harder to control, and more blossoms are ruined, not only by chewing insects, such as Japanese beetles, but also by adverse weather conditions. But avoiding excessive heat and maintaining reasonably uniform amounts of light and air circulation are much easier with outside care, at least in most locations in the temperate zones.

## WATERING AND HUMIDITY

Providing the right degree of water vapor and establishing the right schedule for drenchings are of critical importance for the growth and flowering of vandaceous orchids.

### Relative Humidity

Many growers outside of Asia have mistaken notions about the daytime levels of relative humidity that prevail in the tropical areas where strap-leaf vandas and ascocendas flourish. Contrary to general belief, these are not always steamy places, except during the rainy season. In fact, from midmorning to mid afternoon in the dry season, relative humidity in Chiang Mai is not much different from that in many places in temperate zones in the spring and summer months. To demonstrate this contention, examine the data in Table 8-5, and compare them with your own situation.

Normally, try to keep humidity in the range of 60–80% during most of the daytime. Achieving that may require considerable attention during hot, dry weather and in arid climates. Excessive dryness results in slow growth, desiccation, leaf drop, and poor flower quality. If daytime relative humidity dips into the 45–50% range for a few hours occasionally, that causes no harm. In fact, it helps to harden the leaves, unless the temperature is above 29°C (85°F), in which case action should be taken without much delay if you wish to maintain ideal conditions. Pay particular attention to trying to maintain relative humidity close to 80% on hot, sunny days. Warm air absorbs and holds more water vapor than does cold air. When the temperature in the greenhouse is high, dry air soaks up moisture like a blotter from whatever source it can. Leaves and aerial roots exposed to hot air become desiccated quickly, because their moisture content is higher than that of the surrounding air. Misting the plants and wetting the walkways and floors of a greenhouse can produce water vapor that prevents stress that would otherwise occur. Install misting apparatus low; water vapor rises.

Relative humidity rises at night because of the drop in temperature. Nothing needs to be done to counteract this, as long as air movement is adequate. Misting systems should not be operated at nighttime if humidity goes above 70% without any assistance from equipment.

Table 8-5. Relative humidity and rainfall of Chiang Mai, Bangkok, and Singapore: average of monthly observations.

| | CHIANG MAI | | | BANGKOK | | | SINGAPORE | | |
|---|---|---|---|---|---|---|---|---|---|
| | HUMIDITY % | | RAIN (IN.) | HUMIDITY % | | RAIN (IN.) | HUMIDITY % | | RAIN (IN.) |
| | 6:30 AM | 12:30 PM | | 6:30 AM | 12:30 PM | | 6:30 AM | 12:30 PM | |
| JAN. | 96% | 52% | 0.1" | 91% | 53% | 0.3" | 82% | 78% | 9.9" |
| FEB. | 93 | 44 | 0.4 | 92 | 55 | 0.8 | 77 | 71 | 6.8 |
| MAR. | 88 | 40 | 0.3 | 92 | 56 | 1.4 | 76 | 70 | 7.6 |
| APR. | 88 | 49 | 1.4 | 90 | 58 | 2.3 | 77 | 74 | 7.4 |
| MAY | 90 | 60 | 4.8 | 91 | 64 | 7.8 | 79 | 73 | 6.8 |
| JUNE | 92 | 67 | 4.4 | 90 | 67 | 6.3 | 79 | 73 | 6.8 |
| JULY | 94 | 69 | 8.4 | 91 | 66 | 6.3 | 79 | 72 | 6.7 |
| AUG. | 95 | 73 | 7.6 | 92 | 66 | 6.9 | 78 | 72 | 7.7 |
| SEPT. | 96 | 72 | 9.8 | 94 | 70 | 12.0 | 79 | 72 | 7.0 |
| OCT. | 96 | 69 | 3.7 | 93 | 70 | 8.1 | 78 | 72 | 8.2 |
| NOV. | 96 | 63 | 1.2 | 92 | 65 | 2.6 | 79 | 75 | 10.0 |
| DEC. | 96 | 57 | 0.5 | 91 | 56 | 0.2 | 82 | 78 | 10.1 |
| YEAR | 93% | 60% | 42.5" | 92% | 62% | 55.0" | 79% | 73% | 95.0" |
| NO. OF YEARS | 8 | 8 | 13 | 12 | 12 | 44 | 10 | 10 | 64 |

Source: *Tables of Temperature, Relative Humidity and Precipitation for the World,* Part V, *Asia* 1976. Meteorological Office, Her Majesty's Stationery Office, London.

On days when the weather is cool and overcast, keep humidity near the low end of the 60–80% range, if you can. This may require that you not operate any misting systems and that you raise the setting of the heater thermostat. Cool, damp air promotes rot and fungi. On the other hand, at times when the heating system is operating full blast, the misting system may have to function constantly in order to maintain the minimum humidity as high as 60%, even on dull days.

## Misting Equipment

A number of commercial misting and fogging devices that raise relative humidity (and provide cooling, as well) are on the market. The difference between a mister and a fogger is that the latter atomizes the water to a greater degree. In practice, many foggers really are no more than misters, despite the name given to them by the manufacturer. Some do an excellent job of producing large amounts of fine vapor, but the best ones with enough capacity to be adequate for more than a small greenhouse are very expensive. Many systems do not atomize the water sufficiently; they blast excessive moisture onto some plants and cannot provide uniformly fine mist throughout the greenhouse. Spend ample time exploring the alternatives before making a purchase.

The full benefit of any kind of misting or fogging system can be obtained only if the equipment is properly coupled with fans of adequate size and velocity, strategically positioned to assure uniform dispersion of the moisture after it is emitted. To the maximum extent possible, direct the stream of mist from misting nozzles onto the walkways and the spaces under the benches or hanging plants, not at the plants themselves, and not from above. The mist should not be so heavy, or aimed in such a direction, that any plants are constantly wet, because that will promote rot and fungi.

A simple homemade misting system can be made from oil-burner nozzles connected to pipes or garden hoses. It is inexpensive and does an excellent job of supplying moisture to the air in the greenhouse. Use nozzles with an 80° orifice and an output in the range of about 4 l (1–1.25 gal) per hour, which emit an "extra-solid full-cone spray." These nozzles can be purchased quite inexpensively at establishments that sell oil-burner parts. Any supplier of fuel oil can tell you where to get them. The nozzles come with a fine built-in filter that can easily be cleaned (be sure to clean it periodically). You also will need to buy a special "double-female" adapter, because the thread at the rear of the nozzle is not a normal pipe thread; it is finer. The nozzle thread at the rear screws into one end of the double-female adapter, and the other end of the adapter receives an ordinary pipe-threaded nipple. Any well-stocked hardware or plumbing-supply store can provide the short, threaded, copper or brass nipples and adapters you will need in order to go from the nozzle-adapter to a garden hose or pipe-fitting. (The oil-burner parts firm is unlikely to carry these.) The components are simply screwed together, using Teflon tape (or paste) at the joints.

Humidistats, combined with solenoid valves, can automatically control systems for raising humidity, but supplementary misting by hand generally is beneficial from time to time when solar radiation is strong or outside humidity is very low—automatic equipment may not to be able to provide enough vapor.

### Equipment for Measuring Humidity

Various kinds of hygrometers are available for measuring relative humidity. Mount them out of the path of misters, preferably in or near the middle of the greenhouse or lath house, and at an elevation about the same as that of the average level of the plants. The most accurate hygrometers are the wet-bulb type; however, they are less convenient to read than the dial types because a table must be used to convert the readings into corresponding relative humidity figures. The dial-type hygrometers require adjustment occasionally. I use them, but I check them frequently against a wet-bulb hygrometer.

## Drenchings

High relative humidity does not obviate the need for frequent drenchings of vandas and ascocendas under most conditions. The appropriate schedule depends on the levels of temperature inside and outside the greenhouse and on the volume of air movement around the plants. Thai growers, who grow their plants under lath outdoors, generally water only once a day—on very hot, dry days, some of them water twice. One grower told me that more frequent watering would be beneficial in the dry season, but that labor and water costs are a deterrent.

The lower the greenhouse temperature, the less frequent should be the periodic drenchings of the plants. Under dull, cool conditions, one drenching daily should be ample, and it should be done early in the morning—even that single drenching may not be needed if the relative humidity is higher than 65 or 70%. On rainy days, do no watering at all, and turn off any misters or foggers, unless a desiccating type of heating system requires their use.

Conversely, the higher the greenhouse temperature, the more frequent should be the watering schedule, and the more water vapor should be provided in the intervals between the drenchings. Hot, sunny conditions may mandate several heavy drenchings each day. Give the first as early as possible in the morning. On hot sunny days in arid climates and in greenhouses with large open roof vents, water vapor rises and disappears as fast as it can be produced by most misting or fogging systems. In such situations, there is no substitute for frequent drenchings.

### Drenching Technique and Scheduling

The best technique for watering vandas and ascocendas is to give them a "once over lightly" first, and then wait 5–30 minutes. By the end of the interval, the velamen (the white epidermal covering around the roots) will have become more absorbent. Follow with a thorough drenching until the velamen becomes a uniform grass-green color and the water cascades from the leaf axils down the stem of the plant. The spongy velamen is waxy and sheds water at first. That is why its surface needs to be dampened for a short period before it will absorb much water. Only after the velamen is saturated can it serve as a reservoir of water for the plant. This two-step method is equally important when you are fertilizing (or applying systemic insecticide or fungicide).

The purpose of drenching until water cascades down the stem is to flush out any accumulations of dust or loose vegetative matter lodged in the axils. Such accumulations are a fertile breeding ground for pathogens. Heavy drenchings also rinse residues of fertilizer salts from the leaves—their presence encourages the growth of algae.

Give the first drenching of the day as early as possible, even on sunny days. The plants should first be watered when the temperature is rising; that helps the plants to dry faster, and growth of pathogens is deterred, especially if there is good circulation of air. It also helps to prevent desiccation as the morning temperature rises. If additional waterings are necessary, give the last one no later than about 3:30 in the afternoon, in order that the leaf axils may be dry by nighttime.

A good general guide is that additional drenchings or heavy mistings should not be given as long as the velamen still shows signs of green coloration from the previous drenching. If the velamen takes as long to turn green on any subsequent watering as it did on the first one of the day, the interval between the waterings has been too long.

Long aerial roots that resemble strands of beads also indicate that the intervals between waterings have been too long, and that there has been a prolonged pattern of stop-and-go growth. If plants lose their lower leaves and become leggy, the cause usually is too infrequent watering, although this condition may also be attributable to lack of live roots and aggravated by too low humidity.

### Water Temperature

To avoid shocking the plants, the water should be several degrees warmer than the ambient temperature in the growing area. This practice is vital for very small seedlings; they can readily be shocked by cold water to such an extent that it takes them months before they resume growing. Mature plants take longer to show the ill effects of cold water, but repeated shocks eventually will retard their growth and vigor. Another reason for giving the first heavy drenching early in the morning is that the leaf surfaces of the plants will not yet have been warmed by the sun and are less likely to be chilled when watered.

Even in the summertime, it is highly desirable to temper any water that will hit the plants, regardless of whether the plants are indoors or outdoors. If the plants are hung outdoors, a long length of black garden hose or plastic tubing, strung out so that its surface is exposed to as much sun as possible, may provide ample tempering at zero cost, provided that the periods of drenching are kept short (in order to consume only the water stored in the hose or tubing).

### The Need For Drying Cycles

Vandas and ascocendas, and most of their relatives, require cycles of wetness and dryness during the day (*Paraphalaenopsis* are an exception); however, they do not all have the same optimum drying interval. Thailand's prominent hybridizer,

Charungraks Devahastin, notes that *Vanda coerulea* grows better if it dries off quickly—within one hour after watering. In its native habitat, the wild species grows in bright, exposed locations on scrubby bushes and open trees, where the plants dry quickly and humidity is rather low at times. *Vanda sanderiana* comes from a different environment, where humidity is high throughout the day, and the plants grow on tall trees under a leafy canopy that retards drying after a rain. Most hybrid vandas and ascocendas have a considerable amount of *V. sanderiana* in their ancestry and share some of the latter's need for more humid conditions; they do best if they dry more slowly—Devahastin says over a period of two or, at most, three hours. No evidence of wet leaves or of dampness in the leaf axils should remain for longer than the prescribed drying-off interval. If the plants and their roots do not dry within a maximum of three hours, turn up the heat or increase the movement of air. Stand-by fans, reserved for such occasions, are useful.

The reason for the concern about prompt drying is that evaporation of water from leaf, root, and stalk surfaces, and especially from leaf axils, helps to prevent fungi from thriving and spreading. Fungus spores generally need a film of water or other enveloping moisture if they are to germinate. Rising temperature and good air movement accelerate the process of evaporation. If the aerial roots have been drenched to the saturation point at each watering, as they should be, the velamen will retain enough moisture for the short-term needs of the plants, even though its surface no longer shows any signs of dampness and even if the relative humidity of the surrounding air is rather low.

## AIR CIRCULATION

Good air movement is needed to dry a plant when it is wet, and thereby reduce vulnerability to bacteria and fungi. It also is needed to facilitate the exchange of metabolic gases at the leaf surfaces. Leaves release oxygen and take in carbon dioxide through their pores, or stomata. Stagnant air around the stomata impedes this vital process (so do mineral deposits and other residues that clog the pores); good air movement and clean leaves promote it. In sunny weather, movement of air is needed to dissipate the heat that direct solar radiation produces on the exposed surfaces of the leaves.

Air circulation may mean circulation of an unchanging mass of air or it may mean an exchange of fresh air for an existing mass of air. Most writings on orchid culture stress the importance of air movement, meaning circulation of the air in the greenhouse, and seldom mention the necessity of air exchange. If stale, humid air is recirculated constantly, no amount of velocity will keep the air healthy. Pathogens will merely go around and around, and sooner or later they will cause disease.

Do not try to make your greenhouse absolutely airtight in wintertime by enveloping it with plastic film. Be sure your orchids are getting some fresh air, preferably a small amount constantly, even in midwinter.

Flow of air around all parts of the roots and leaves of vandaceous orchids cannot be assured if there is overcrowding, so it is important to place plants far enough apart that circulation of air around them is unimpeded. Make bench supports as thin as possible, so they will not hamper circulation, and do not use under-bench areas for storage purposes—leave the areas open, in order to facilitate circulation. Good growing conditions are airy growing conditions.

It is tempting to try to achieve adequate circulation of air around crowded plants by increasing the velocity of air movement, but that is not without risk. If the air is low in humidity, high-velocity air circulation has a desiccating effect. If the air is damp, and the temperature is not kept reasonably high, high-velocity air has a chilling effect around the roots and leaves that stunts vegetative growth and flower quality.

## Improving Air Circulation with Fans

In addition to moderating temperature and distributing moisture, fans are essential for regulating the amount and patterns of air circulation. By preventing stagnant air from hovering about the leaves of plants, they promote metabolic gaseous exchange and a healthy growing environment. Most growers do not use enough fans and do not put enough thought into their selection and placement. That is a shame, because few actions a grower can take can yield as beneficial long-term results.

Place some fans in locations and at angles that will result in the air being moved up or down. Position other fans to move air laterally around all sides of all of the plants, rather than just around some of them. Oscillating fans are excellent, but seldom should be used exclusively. Some growers use powerful fans to send a blast of air down the walkways. While that produces a certain amount of general turbulence, the overall benefits may be very limited; if the plants are on steplike benches (and especially if the benches are covered with a solid material), the plants nearest the walkways and on the lowest bench may receive adequate or, more likely, excessive air movement, but those on the upper and back benches or sides of the greenhouse may receive little. Uneven air movement of that sort creates substantial differences in microclimates, and hence in the vigor of some plants and in the quality of their flowers. The differences in microclimates are even more pronounced if moisture from misters and foggers is not well distributed throughout the house.

### Testing Air Circulation

There is a very simple and inexpensive way to test the patterns and intensity of the circulation of air in a greenhouse or a lathhouse. Buy about a dozen votive candles (the short, squat kind that can stand on their own base without falling over). Place them in key locations, laterally and vertically, throughout the growing area. Avoid rearranging any plants, because that might distort the findings. Light all of the candles. The orientation of each flame indicates the direction of the air current. The angle indicates the force of the current, as does the speed with which the candle burns. Any candle with an unflickering, upright flame is in a dead spot.

Experiment with the number and arrangement of various types of fans, and change their angle of inclination and orientation until you create the desired patterns of air flow. You may also need to rearrange plants if large specimens or crowded benches block some of the air currents. Optimum arrangement changes with the seasons, and testing of the results must be repeated. Bear in mind that in a greenhouse with mixed genera, it may be essential to create differences among the microclimates, rather than uniformity. The differences, however, should be of the grower's choosing and not simply accidental.

# 9

# Nutrition and Other Aspects
# of Plant Care

## GENERAL PRINCIPLES

If you follow the recommendations for light, temperature, water, and air circulation, you are far along the way to growing your vandaceous orchids well. Fertilizers will keep healthy stock growing well, and chemicals may prevent or cure disease, but they are no substitute for the basics. While there is more than one horticultural road to success, all of the regimens described in this chapter are based on common principles.

Vandas, ascocendas, and their relatives should be fertilized frequently; they are heavy feeders. They have no pseudobulbs in which to accumulate large reserves of food, and their aerial roots do not have the benefit of a surrounding medium that gradually releases dissolved mineral nutrients. Excessive dosage of mineral nutrients, however, can do much more harm than low dosage.

Expert growers use blends of fertilizer that are lower in nitrogen and higher in phosphorus and potassium at times when their plants are putting forth inflorescences and flower buds. All of them make considerable use of high-phosphorus fertilizers throughout the year, even though frequency of use varies seasonally. Excessive application of nitrogen produces soft, tender leaves that are prone to fungi and bacterial rot. A healthy plant has firm, stiff leaves.

The use of vitamin-hormone additives is widespread in the United States, as is the occasional addition of mineral supplements or trace elements, whereas those seem to be rare practices in Thailand. The cost of such additives is minimal; if in doubt, use them. But be sure to use such supplements with great moderation; do not overdo them. Plants in the *Vanda* alliance have a low tolerance of excessively frequent or excessively concentrated applications of fertilizers, trace elements, and hormones. The sensitivity is exacerbated when there is no potting medium to act as a partial buffer.

A vital question for every grower is: "Are my plants getting the optimum amount of each of the major nutrients and of all the needed trace elements?" If your plants are growing vigorously, blooming frequently, and producing flowers of good quality and size, your nutrition program is right on the mark. If you still have doubts, or if your plants aren't doing as well as you think they ought to be, consider having leaf-tissue analysis performed.

Nutrition, in some instances, can explain the difference between a superbly grown and flowered plant and one that fails to reach its potential. It also can affect the capability of a plant to resist pests and diseases. That is why it requires careful attention even though there are no precise guidelines to follow. Trial and error, within the boundaries described, combined with the use of a variety of fertilizers, is the best you can do.

Water quality is another important consideration. The ability of fertilizer to work to the maximum advantage of the plants depends partly on the quality of the water in which it is dissolved. But beyond that, some water supplies contain chemicals and concentrations of minerals in amounts that can cause symptoms of toxicity in plants.

Pests and diseases usually are not a major problem for most growers of vandas and ascocendas, but prophylactic measures to prevent buildup of pathogens can be helpful. Most important in this regard is to keep the growing area completely clean and free of all detritus.

### Cautions

The opinions and suggestions regarding cultural practices and/or specific products in this chapter, as well as in the rest of the book, are based on my personal experience, under my conditions, or are based on the experience of other growers whose practices I have observed. Further experience may cause me to change my opinion, or may reveal risks or cautions associated with the fertilizers, bactericides, fungicides, algicides, disinfectants, and other chemicals discussed here. Conditions of other growers under other circumstances may differ materially and may adversely affect results.

If you apply any of the products, or follow any of the practices in your own circumstances, do so cautiously. Carefully read the instructions and other information supplied with the product. If you have any doubts or questions, seek the current advice of the manufacturer and of appropriate governmental agencies before you use the product. Nearly all manufacturers have technical staff to assist customers. Usually either the product label or the retail supplier can provide a telephone number or address of the manufacturer. And do not hesitate to ask experienced growers about their experiences; anecdotal evidence is not scientific, but it can at times be valuable.

## REGIMENS FOR PROPER NUTRITION

### Fertilizers and Trace Elements

Just as there are considerable individual differences in the dietary habits of orchid growers themselves, so do growers follow a variety of regimens when it comes to nourishing their orchids. Orchid growers tend to be faddists when it comes to choice of fertilizers, nutrition supplements, and potting media. For a while, they tout one brand or formula of fertilizer, or a certain nutritive supplement; then, a year or two later, they just as enthusiastically describe the improvement they have witnessed since changing to some other brand, formula, or additive. The process never seems to end. Unfortunately, we never really know whether we have achieved the optimum regimen. The "best" regimen for fertilizing plants can be "best" only in relation to a number of other relevant considerations, such as light intensity, length of day, temperature patterns, and other elements that are forever changing. This reality, combined with the gullibility of anxious orchid growers, is a great boon to sellers of fertilizers and various vitamins, hormones, mineral supplements, and assorted nostrums.

Fertilizer formulas vary in the amounts of nitrogen, phosphorus, and potassium supplied, as well as in the degree of acidity of the product. The number and quantity of trace elements in fertilizer blends also varies, depending on the manufacturer and the product. Read the label on the package; some manufacturers provide more information than do others.

Nitrogen promotes vegetative growth; phosphorus is vital for a plant's respiration and for production of blossoms; and potassium plays an important role in plant metabolism and root growth. The sources of nitrogen differ, and in some situations the type of nitrogen may be significant. For plants in the *Vanda* alliance, most of the nitrogen should come from nitrates, because nitrogen from that source is more readily available to aerial roots; ammonium and urea forms require warmth and bacterial action to convert them to nitrates before the nitrogen becomes available to the plants. Most fertilizer packages disclose the percentage of nitrogen from each source.

Practically all fertilizers commonly used by orchid growers contain some (and most likely sufficient) trace elements. The trace-element contents frequently are not stated on the package because some states have regulatory restrictions regarding minimum concentrations mandated for listing. These minimums are based on field crop requirements, which often are excessive for most horticultural purposes.

All water supplies contain some trace elements. Be very careful about adding more than already are in whatever fertilizers you are using, unless you have some evidence of need; more is not necessarily better—in fact, it may well be harmful. Be cautious about fertilizers that claim to have an exceptionally high trace element content. If the claim is true, the higher levels of trace elements may harm some of your more tender plants, especially if the concentrations of certain trace elements in your water supply are above normal (a good reason for having your water analyzed). For example, iron, zinc, and boron are known to cause symptoms of toxicity when in excessive supply (Withner 1974).

Some manufacturers endeavor to justify substantially higher prices than those of their competitors by claiming that their fertilizers are effective at smaller dosages. What determines cost-effectiveness is the cost per quantity of parts per million (ppm) of nitrogen, phosphorous, and potassium (and of the trace elements). Consider, for example, a 7-7-7 blend and a competing 21-21-21 blend. Assume that both blends have a satisfactory composition of trace elements and are water soluble, which is the case with major brands. A teaspoon of the 21-21-21 blend produces the same number of ppm of nitrogen, phosphorus, and potassium in solution as does a tablespoon of the former (or 5 cc produces the same ppm as 15 cc); consequently, for cost-effective equivalency, the 7-7-7 blend should be priced at roughly one-third the level of the 21-21-21 blend, but in practice there rarely is a difference.

Sometimes a manufacturer will tout the efficacy of a specific trace element for orchids, when in fact the background research is questionable and may pertain not at all to orchids, but rather to other, unrelated plants. Silicon, for example, is heavily advertised as an additive to protect orchids from fungal infection, but the main research on which this claim is based was done with cucumbers grown hydroponically in a recirculating solution. The reported enhanced protection against fungal infection referred to mildew, which seldom is a problem for orchid growers. Silicon is the second most common element in the earth's crust and is found in most water supplies. It is not in short supply. No scientific analysis, to my knowledge, has been conducted to determine the optimum or normal concentration of silicon in dried tissue of orchids, either generally or for any individual species. Using a silicon additive in a fertilizing program for orchids, therefore, is a matter of faith (or gullibility), rather than a scientifically established benefit.

## Nutrition Regimens of Successful Growers

Orchid growers yearn to learn the specific recipes and practices of successful growers, as if there were some magic formula. There is none. A review of the fertilizing regimens of some of the best growers of vandas and ascocendas in this country and in Thailand may, however, be helpful.

### Regimens of Two American Growers

One outstanding grower is Robert F. Fuchs of R. F. Orchids, located in Homestead in southern Florida. Fuchs has done more than anyone else to make high-quality vandas and ascocendas readily available in this country and to demonstrate that they can easily be grown in all parts of the United States, if simple instructions are followed. His credentials as a successful grower and hybridizer are attested to by the fact that he has won by far the largest number of awards from the American Orchid Society for the quality of his vandaceous plants. His regimen for fertilizing is succinctly stated (Fuchs 1990). It is worth repeating verbatim.

> At our nursery, we use Peters 20-20-20 (All Purpose) fertilizer weekly during the growing season. Inside the greenhouse, under a controlled environment, the concentration we use is the standard recommended one of 1 teaspoon per gallon (13 ml per 10 liters). Outside, we increase the concentration to 1 tablespoon per gallon (40 ml per 10 liters). . . During the winter (i.e., the non-growing season) we apply the same proportions every two weeks rather than weekly. In addition, at every third feeding, we substitute Peters 10-30-20 (Blossom Booster). This substitution applies to plants inside or outdoors, and to all seasons of the year. Further, once a month, we add ¼ teaspoon (3 ml per 10 liters) of SUPERthrive, a concentrated vitamin and hormone solution, to each gallon of fertilizer solution.

Another superb grower is Marilyn Mirro, who grows vandas and ascocendas to perfection on Long Island, New York. She is widely recognized as the premier grower of these two genera in the northeastern United States. Her regimen (Mirro 1990) is as follows.

> The amount of fertilizer depends, of course, on the time of year . . . The different fertilizers are rotated. From October through January, I use ¼ teaspoon of Peters 20-20-20, or alternatively 10-30-20, per gallon of water (3 ml per 10 liters), with a drop of SUPERthrive added to either. From February to June 1, I increase it to ½ teaspoon of Peters per gallon of water (6 ml per 10 liters). The community pots and seedlings are fertilized with a minimum rate of ¼ teaspoon per gallon (3 ml per 10 liters) year-round. In the greenhouse, the vandas are watered first and then fertilized on every sunny day. In a week of sunny days, for example, I'll start out the first two days with a solution of 20-20-20, and, on the third day, I'll use Peters 10-30-20. On the following day, I'll use liquid seaweed at the rate of one tablespoon per gallon (40 ml per 10 liters). During the winter months, I'll cut back to once a month. In my opinion, liquid seaweed provides the plants with trace elements that are absent in chemical fertilizers . . . When the plants are outdoors from June to September 1, I increase the dilution rate to one teaspoon of Peters per gallon of water (13 ml per 10 liters). During the summer I also use Sequestrene iron at ¼ teaspoon to the gallon (3 ml per 10 liters) once a month, as well as magnesium sulfate (Epsom salts) at the rate of one tablespoon per gallon (40 ml per 10 liters) every two weeks.

## Regimens of Thai Growers

One of the best Thai growers, located in Chiang Mai in northern Thailand, uses a 21-21-21 fertilizer weekly, at a rate of 13 ml per 10 liters (1 tsp per gal) of water, but in the rainy season (May through September), he uses a lower-nitrogen formula. For community pots and very small seedlings, he applies a 3 ml per 10 liters (0.25 tsp per gal) solution daily. When the flower buds begin to swell, he shifts to a high-phosphorus fertilizer to make the flowers larger and the colors deeper. He also said that, at that time, he uses a glucose solution of 13 ml per 10 liters (1 tsp per gal) of water, which he asserted is "a substitute for photosynthesis." His vandas and ascocendas were superbly grown, and flower quality was outstanding. He applies fertilizer solutions as a fine spray.

A prominent Bangkok-area grower also uses glucose on seedlings, at a rate of 1 ml per 10 liters (4 tbs per 100 gallons) of water weekly, along with a high-potassium fertilizer. That is a much weaker dosage of glucose than that used by the Chiang Mai grower. For more on glucose, see "Use of Glucose" in this chapter.

Another outstanding grower in Chiang Mai uses a 30-10-10 fertilizer alternated with a 10-30-20, or 13-27-27, at a rate of 25 ml per 10 liters (2 tsp per gal), with greater emphasis on one of the low-nitrogen formulas just before the plants begin to send out new inflorescences and during their blooming period. He applies fertilizer twice weekly. He is one of only two growers I encountered who said they use 30-10-10. The others seemed to be chary about using strong concentrations of nitrogen. The second grower who uses a 30-10-10 blend resides in Bangkok and applies it weekly in the sunny season. In the rainy season, he shifts to a 10-30-10 formula.

Two Bangkok growers say they get darker-colored flowers if they start to hold back water and fertilizer when the flower buds are well formed; they merely mist lightly and do not feed. Dehydration does cause increased anthocyanins, but the margin before damage is done to the plant is very small, so if you follow their practice, do so with great caution. Both growers maintain that nitrogen must be used minimally, because it tends to produce lighter-colored flowers. Both also recommend a routine of using different brands and formulas of fertilizer, on the grounds that the quantity and proportions of trace elements differ, and that these have a bearing on flower color. They recognize that a more scientific approach would be preferable, but that is not possible, given our current state of knowledge.

## My Personal Regimen

I am very skeptical about any asserted superiority of one brand of fertilizer over others; what matters are the specific formulas and the degree of solubility of the mineral nutrients (which is visible to the user). By alternating a number of different fertilizers, I, like some of the Thai growers, believe that from one source or another my plants will get all the nutrients and trace elements they need, and will not get any single element in a toxic amount. All the formulas I use have ample potassium as well as phosphorus. The pH of my water supply is 7.8, with a low bicarbonate content.

I have used a 20-20-20 formula interspersed with a variety of other formulas, two of which are very high in phosphorus (30% and 50% respectively). From May to September, I use one of the latter two blends about every fourth time, at a rate of 13 ml per 10 liters (1 tsp per gal). I feed about every third day, if it is sunny and warm. (I moisten the plants first, but lightly.) The high phosphorus content stimulates abundant production of new roots and good flower color.

Recently, instead of a 20-20-20 general-purpose, I have been using a 5-11-26 blend designed for hydroponic culture, because, in effect, my vandas are being grown "hydroponically," since I use no potting medium. All of the nitrogen in the blend comes from a nitrate source, which is an advantage, and the formula is slightly richer in most trace elements than are some other blends. About every sixth fertilizing, I use calcium nitrate to supply additional nitrogen and calcium (my water supply contains little calcium).

Seasonal and climatic considerations are taken into account with respect both to dosage and to the nitrogen-phosphorus content of the fertilizers I choose. In the wintertime, I use only 6 ml of fertilizer per 10 liters of water (0.5 tsp per gal), about every fourth or fifth day (I do not fertilize on dark or rainy days). I also reduce the total application of nitrogen, by using blends with a lower content of nitrogen.

Occasionally, I use SUPERthrive,* a vitamin-hormone supplement, at the recommended rate of a drop or two per 4 l(1 gal) of water. I have been unable to find any reliable research to support its efficacy, but there is considerable anecdotal support from growers who use it. I remain skeptical, and I am not sure my plants need it any more than I am convinced that my wife is right when she insists I take a daily vitamin pill, but I feel it cannot do me any harm, and it may possibly do me some good. Some excellent growers never use vitamin-hormone supplements. I regard my usage as a kind of low-cost insurance, rather than a documented benefit.

### Use of Glucose

The aim of the Thai growers who apply glucose directly to the roots and leaves of their vandas and ascocendas is to give the plants a nutritional boost by supplementing the glucose formed by photosynthesis. Glucose is especially beneficial when a plant has been badly stressed and its leaves have lost their turgidity—for example, because of excessive heat and insufficient moisture—because the ability of the leaves to create glucose by photosynthesis has been impaired.

So much for the rationale. But does it work in practice? I tested the procedure

---

*Mention of trademark, proprietary product, or vendor does not constitute a guarantee or warranty of the product by the publisher or author and does not imply its approval to the exclusion of other products or vendors.

on some badly stressed plants and was astonished by the favorable results, so much so that I now regularly "sweeten" my vandas and ascocendas about every three weeks, as part of my normal feeding schedule. It makes the leaves plumper. For convenience' sake, however, I use ordinary household sugar instead of glucose. Household sugar (cane or beet) is a carbohydrate composed of glucose bonded to fructose, so it is a more complex sugar than is glucose. Apparently, the plants can extract the glucose by metabolism, and they may possibly utilize the fructose molecules as well. In any event, the sugar solution appears to give excellent results. Sucrose, incidentally, not glucose, is the major ingredient (apart from water) in the original flasking formula of Lewis Knudson, the pioneer of asymbiotic culture of orchid seedlings (culture without dependence on symbiotic fungi), and it still is the usual sugar used in nutrient mediums in laboratories. Since it is beneficial in promoting growth of seedlings in flasks, it seems reasonable that it would help older plants as well.

I make a solution of equal parts of sugar and a water-soluble fertilizer, at whatever dosage of fertilizer I customarily use at that time of year, and spray it onto the roots and leaves of the plants until the velamen of the roots has turned a bright green. As with other applications of fertilizer, I first dampen the roots slightly.

Sugar may be a tonic for some fungi as well. Moreover, it may attract ants. In order to control these side effects, I drench the entire greenhouse the following day with a detergent (Physan or its equivalent) to wash off any residue of sugar. By following this procedure, I have had no adverse side effects. A word of caution, however. I do not recommend using a sugar solution on any plants potted in a medium, because it would be much more difficult to control fungi and a breaking-down of the potting medium. Similarly, I stop using the sugar solution whenever I see any signs of a fungus problem and do not resume it until the problem has been corrected. This situation, however, has seldom happened.

### Leaf Tissue Analysis to Detect Inadequate Nutrition

Leaf tissue analysis may help identify a nutritional problem. Table 9-1 shows values for a sizable batch of representative leaves from the plants of each of the three American growers whose feeding regimens are described in this chapter. All of the leaves were from well-grown vandas. Most growers probably will be satisfied if their plants are as well nourished as those of the three growers cited.

In reading the table, understand that the highest values for each element are not necessarily "better" than the lowest ones; the differences may be due to the timing of the most recent application of fertilizer or to trace elements in the water supplies. The vandas and ascocendas of each of the three growers have won numerous awards for their quality in competition with the best plants of other growers

Table 9-1. Tissue analysis of *Vanda* leaves of three growers[a].

| VALUES IN PERCENT (%) | | | |
| --- | --- | --- | --- |
| ELEMENT | | GROVE | MIRRO | R. F. ORCHIDS |
| Nitrogen | (N) | 1.1 | 0.7 | 0.9 |
| Phosphorous | (P) | 0.11 | <0.1 | <0.1 |
| Potassium | (K) | 2.34 | 2.29 | 1.59 |
| Calcium | (Ca) | 2.85 | 2.94 | 2.33 |
| Magnesium | (Mg) | 0.250 | 0.089 | 0.168 |
| Sulfur | (S) | 0.052 | 0.087 | 0.125 |

| VALUES IN PARTS PER MILLION (PPM) | | | |
| --- | --- | --- | --- |
| ELEMENT | | GROVE | MIRRO | R. F. ORCHIDS |
| Iron | (Fe) | 107.0 | 105.0 | 58.3 |
| Manganese | (Mn) | 73.4 | 226.0 | 102.0 |
| Zinc | (Zn) | 15.6 | 19.0 | 9.80 |
| Copper | (Cu) | 2.07 | 2.01 | 1.11 |
| Boron | (B) | 17.7 | 18.1 | 9.66 |
| Molybdenum | (Mo) | 0.705 | 0.729 | 0.384 |
| Aluminum | (Al) | 25.8 | 26.5 | 13.7 |

[a] Leaf tissue analysis performed by Fisons Analytical Laboratories, Warwick, New York, July 1991.

nationwide. That suggests that their plants are neither lacking in any important nutritive element nor suffering from any overdosage. The lowest readings may safely be regarded as adequate, and the highest may be regarded as not excessive. Unfortunately, there has been no extensive research on the mineral nutrient content of the tissue of vandas and their relatives. Research on some other genera of orchids is of limited relevance, because substantial differences have been found one genus to another.

For most growers of vandaceous orchids, neither nutritional deficiency nor water pH is likely to explain really poor vegetative growth or greatly inferior flower quality. Seldom is a plant's survival threatened by the fertilizing regimen unless the dosages are excessive. Instead, the climatic conditions of light, temperature, moisture, and air circulation are the most likely culprits requiring corrective action.

## WATER QUALITY

Water of good quality is necessary in order to grow healthy, vigorous orchids of any sort. Agricultural or industrial pollutants in the water supply, heavy concentrations of certain naturally occurring minerals, and the pH of the water source are the main concerns. Every grower should obtain a detailed analysis of the water he or she uses for misting and drenching. The analysis should specify the normal range of each element, as well as providing the concentration found in your water sample. Local Cooperative Extension Service offices can tell you where testing for horticultural purposes can be performed—usually it will be at the horticultural department of your state university, and the charge will be small.

## Pollutants and Minerals in the Water Supply

In some agricultural areas, the water supply is so heavily loaded with residues of fertilizers, herbicides, and other agricultural chemicals that the water is toxic to orchids if used constantly and exclusively over an extended period. A comparable situation is found in some industrial areas.

In some areas, concentrations of such minerals as sodium, iron, sulfur, zinc, and calcium may be a problem. The problem may be one of aesthetics, of toxicity, or of both. For example, a high content of calcium will leave unsightly deposits on plant foliage but, unless extraordinarily high, will not be toxic, whereas a heavy concentration of iron may not only leave deposits on the leaves, but its absorption by the roots may also be somewhat toxic. A high concentration of sodium most certainly will be toxic.

Unsightly mineral deposits on leaves should be removed from time to time, not only for the sake of appearances but also because they may block the pores of the leaves and interfere with respiration. They also block sunlight and thereby reduce photosynthesis. Adding a wetting agent (not a spreader-sticker) to the irrigation water may help to minimize the accumulation of such deposits. AguaGro 2000-L is one such wetting agent.

Problems of toxicity require more drastic and complicated action. The only practical way to deal with the problem of toxic pollutants and excessive concentrations of minerals is by collecting rainwater or by filtering the water by a process called reverse osmosis. If a problem of mineral pollution is not too severe, some untreated water may be diluted with the better water and used. This may be unavoidable if rainwater is scarce and treated water limited. Large open-top galvanized steel tanks of the sort used for providing water for cattle are available at many farm supply stores and are not expensive; they are excellent for storing rainwater or filtered water. Paint the inside of the tank with a not-toxic paint (consult the Cooperative Extension Service for advice), because the zinc coating on the tank might cause some toxicity otherwise.

## Water pH

The subject of the pH of the water source often receives much attention among orchid growers in their discussions with one another. In fact, it receives more attention than it deserves. In and of itself, it seldom creates much of an impediment to successful culture. It is easy to remedy either by adding an offsetting chemical to the irrigation water or by choice of fertilizers with an appropriate degree of acidity. If you are having trouble growing your plants in the *Vanda* alliance well, look first to the environmental factors described in the preceding chapter, and also to undue concentrations of minerals in the water supply. The chances are high that this is where the problem really lies, rather than in the pH of the water supply.

The main reason why the pH of irrigating water is widely viewed as important is that the water solubility of mineral nutrients (e.g., nitrogen, phosphorus, potassium, and the many trace elements) varies with the level of the water's pH, and solubility determines the ability of a plant to absorb the mineral nutrients from the fertilizer solution. The range of pH within which a specific nutrient is maximally absorbable and available to plants differs from one nutrient to another, and is affected by the buffering qualities (i.e., bicarbonate content) of the potting medium, if any is used.

For orchids, the optimal pH of water used for irrigation generally is considered to lie in the 5.8–6.5 span, which is slightly acidic. There is nothing crucial about these limits. Many growers do very well with water that is beyond the upper end of that range. For example, Thomas J. Sheehan (1990), an eminent orchid authority and plant scientist at the University of Florida, states that he has used city water with "chlorine, fluoride, and a pH of 8.5 . . . directly from the tap for 35 years with no ill effects." Some research indicates that the growth of some harmful fungi (e.g., *Fusarium oxysporum*) is inhibited at pH levels above 7.0 (Engelhard *et al* 1989).

## Definition of pH

Water molecules ionize into positive and negative ions—a positively charged hydrogen ion ($H^+$) and a negatively charged hydroxide ion ($OH^-$). An acid is a substance that gives off hydrogen ions when it is dissolved in water. By so doing, it increases the concentration of hydrogen ions in the solution. A base is a substance that accepts hydrogen ions when it is dissolved in water. Since water can both accept and give off hydrogen ions, it may be said to be both an acid and a base. Pure water has an equal number of hydrogen and hydroxide ions.

The pH scale, which ranges from 1 to 14, measures the degree of concentration of hydrogen ions, and hence the degree of acidity. The scale is a negative logarithmic function; a decrease of a single unit on the scale represents a ten-fold increase in the concentration of hydrogen ions. For example, a pH of 4 is 10 times as acidic as a reading of 5, and 100 times as acidic as a pH of 6.

The higher the concentration of hydrogen ions, the lower the pH, and the greater the acidity. The greater the concentration of hydroxide ions, the lower the concentration of hydrogen ions; hence, the higher is the pH and the more basic is the solution. A pH of 7.0, midway on the scale, indicates a neutral solution, which, like pure water, contains a like amount of positively charged hydrogen ions and of negatively charged hydroxide ions.

The terms "pH" and "alkalinity" often are used as if they were synonymous. They are not. While pH is a measure of the concentration of hydrogen ions in a solution, alkalinity is a measure of ability to neutralize acids. The degree of alkalinity is expressed by the number of parts per million (ppm) of bicarbonate in the solution. A standard analysis of the water used for irrigation will provide both measurements. For most horticultural purposes, the desirable level of alkalinity is in the 0–100 ppm range; the normal range is 0–180 ppm. The higher the alkalinity of the irrigation water, the greater will be the buffering (i.e., raising of the pH) of the fertilizer solution and of the potting medium.

It is surprising how few orchid growers understand the meaning of the "calcium carbonate equivalent" measure of acidity that sometimes is printed on fertilizer packages, and its relevance for modifying the pH of the water used. The figure does not indicate the calcium carbonate (i.e., lime) content of the fertilizer. If it did mean that, quite obviously, the higher the figure, the more basic the blend would be. Instead, it indicates the amount of calcium carbonate that would have to be *added* to a ton of the fertilizer in order to raise the pH to 7.0, or neutral. Nearly all commercial blends are on the acid side, which is to say that they have a positive calcium carbonate equivalent rating. Those with a rating of, say, roughly about 400 kg per metric ton (900 pounds per ton) may be considered to be especially acidic, and therefore especially useful in counteracting a high pH or high alkalinity of the water used as the vehicle for the fertilizer. What is most important is the pH of the fertilizer solution, rather than that of the water itself.

## Consequences and Treatment of High pH

If a solution used for fertilizing has a pH of more than 6.5, the minerals phosphorus, manganese, boron, zinc, copper, potassium, and magnesium are somewhat less readily available to plants than they are at a pH in the range of 5.8–6.5. Above pH 7.0, the availability of iron begins to diminish gradually, while that of phosphorus, manganese, boron, copper, and zinc falls rapidly. All this can make a difference if the quantities of these elements are barely marginal both in the fertilizer and in the water used to dissolve fertilizer. That marginality is not always the situation, and some authorities say that it seldom is the case.

For a grower whose water supply has a high pH (and especially if its alkalinity is high), there are two easy courses of action worth considering. One is to lower the pH of the water by adding enough phosphoric acid to reduce it to within the

5.8–6.5 range. I believe that is the best practice. The greater the alkalinity of the water, the more phosphoric acid will be needed to lower the pH. The other possible course of action is periodically to use fertilizers that are more acidic than most—in other words, ones with a high calcium carbonate equivalency—and which also contain an adequate amount of trace elements. A number of these fertilizers are on the market.

### Consequences and Treatment of Low pH

If the water supply has a pH of less than 5.8 but above 5.0, the availability of most mineral nutrients may be very slightly reduced, but not enough to be of any great concern. Certain elements, including calcium, potassium, and sulfur, are less soluble in water of low pH. A generous and varied feeding program can compensate for any minor reduction in the relatively limited availability of these elements. Use a fertilizer with a low calcium carbonate equivalency from time to time, or supply calcium by applying calcium nitrate occasionally, at a rate of about 6 cc per 10 liters (0.5 tsp per gal) of hose-end output. Never mix calcium nitrate with other fertilizers, to avoid precipitation of any phosphorous content they may have; however, fully diluted solutions can be combined. Excessively acid water also can be treated by using a solution of sodium hydroxide or potassium hydroxide to raise the pH.

## CONTAINERS AND POTTING MEDIA

Many growers of vandaceous orchids prefer to grow their plants in slatted wooden baskets with no medium in them. That is the traditional way of growing the plants in Thailand, although some growers there are shifting to perforated clay pots (with no medium). Wood baskets provide excellent surfaces for roots. The surface fibers of the slats absorb some water and stay moist for a short time after the plants have been drenched. Young roots attach themselves to the slats and grow and branch along them, much as they would on a tree limb in the wild. Air can move freely through the slats around the base of the plant, which helps prevent the inception of stem rot.

Clay pots with perforated sides for aeration may also be used. No medium is needed, but the pots must be completely saturated each time the plants are watered; the pots will then gradually release moisture to the roots, and the roots will attach themselves to the pot surface just as they do to wooden slats.

Plastic pots sometimes are used for vandas and ascocendas. Their surfaces are not water-retentive, so the pots, if any roots are to grow within them, must contain a medium, such as very coarse bark chips, tree-fern fiber or chunks, or good-sized pieces of hardwood charcoal, to provide some residual moisture after watering. It is highly desirable that the pots have a number of holes cut into their sides in order to assure adequate ventilation around the base of the plant.

## Wooden Baskets

Vandas and ascocendas traditionally have been grown in slatted teak baskets because most growers regard them as superior to pots of any kind, in most circumstances. Teak, an excellent, rot-resistant wood, was available and inexpensive in Thailand. The baskets provide an ideal combination of moisture retention for a short period and exposure of the roots to air, and they permit aerial roots to develop freely. The evaporation from the slats after drenching provides humidity around any attached roots and encourages multiple small roots, which are better than a few long ones of greater diameter.

The amount of surface area on the slats is surprising. A typical 10 cm (4 in) square-shaped *Vanda* basket has approximately 774 sq cm (120 sq in) of exposed surface to which roots can attach themselves and derive moisture. A 15-cm (6-in) basket has about 1,160 sq cm (180 sq in).

Some growers do use a medium in the baskets. Since most of the roots eventually thrust through the slats and cannot be confined to the medium, it really serves no purpose, except for very young plants with short root structures that may benefit from a surrounding substance in the early stages of their growth. A medium can also help to stabilize the position of newly potted seedlings, but cellophane twine can be used to tie the base of a seedling to the floor slats and hold it just as securely as any potting material. Seedlings should be anchored no matter what kind of container is used, in order to encourage attachment of roots to its surfaces.

## Managing Protruding Roots

It is not essential to manipulate protruding roots back into the basket. They can be left to grow like long tresses; however, they will not branch as much and eventually will become very long—60–90 cm (2–3 ft) or even more. That makes the plant unwieldy to exhibit in a show or to display in your home. In the greenhouse, it usually leaves no space under the plants for growing seedlings or anything else— a drawback in some circumstances. Finally, it becomes bothersome or impossible to pass under the plants. When you are searching for a particular plant or wish to inspect one, it can be like trying to thread your way through a jungle. Nonetheless, vandas and ascocendas can be grown perfectly well in this manner; in fact, from the standpoint of good culture, that probably is the best way to grow them. If the greenhouse has considerably more height than normally found in hobby greenhouses, dangling roots are less of a problem. Despite the inconveniences, I grow all my vandaceous orchids this way.

If you wish to manipulate protruding roots back into the basket, be sure the roots are thoroughly wet to their core before attempting to bend them; otherwise they will certainly crack or break off. While mere cracking of the velamen usually will not cause the outlying portion of the root to die, it is best to avoid the damage. The surest way to saturate the roots is to soak them in a bucket of warm water for about ten minutes, but a very thorough drenching or misting also can accomplish this purpose. When bending, make your movements slow and gentle, and try to avoid making any sharp angles, especially if the root is thick; otherwise the filament at the core will be severed. That fibrous filament is the means by which moisture and nutrients pass from the velamen covering and root tip to the rest of the plant. Normally, the remainder of the broken root will send out new root buds along its length, or at least near the fractured end, but that takes time.

Most amateur growers who let the roots dangle eventually perform some pruning of the longest roots, because of the nuisances already mentioned. In one way, moderate pruning is beneficial, because it will force the roots to branch. In another way, however, the result is not so good. Pruning temporarily reduces the vigor of the plant while it is replacing the lost mass, and that may adversely affect flower quality for a year or two. I try to keep pruning to a bare minimum, even though it does require me to relocate some plants with very long roots from time to time.

If long roots have become such a nuisance you feel you must prune them, do so only during a time of very active root growth; namely, in the spring or early summer. Never do it at any other time. The risk of retarding growth, or even of shocking the plant into a decline from which it will not recover, is too great to take. Instead of pruning all of the roots at the same time, do it over a period of several weeks. Wait until the cut roots have begun to send out new root buds before performing further pruning. Always sterilize any cutting tool over an open flame, or soak it between cuts in a strong solution of trisodium phosphate. Use a high-phosphorus, high-potassium fertilizer liberally for several weeks after pruning.

Instead of pruning the long roots, you might surmise that a sensible procedure would be to form them into a hank or bundle and then wrap that around the outside of the basket, tying the hank to the slats at several points. I tried that with a substantial number of plants one year, with uniformly bad results. The coiled roots, even though not coiled tightly, never sent out many new root buds; most of the old root mass died. I lost more plants than I like to recall, and others were set back considerably. Apparently plants respond poorly not only to any marked crowding of their roots but also to any altered orientation of their roots relative to the horizontal (probably due to adverse effects on the hormonal flows). Pruning the roots is better than forming bundles for attachment to the outside of the basket or shoving them inside the basket, even if it is large enough to permit that. A few individual roots can safely be tied to the outside of the basket, but leave space between them. Best of all is not to disturb them, and thereby leave their orientation unaltered.

## Clay Pots

Clay pots vary in their rate of absorption of moisture, depending on the quality of clay used, the method of firing, and whether they were treated with silicon to inhibit algae. By far the best are Georgia clay pots with slits on their sides. These permit good ventilation and are sufficiently absorbent to retain some moisture. I prefer the shallow style, called pan-pots. With two or three drenchings on sunny days, these pots can be used very satisfactorily in many situations without any potting medium.

Clay pots, if well saturated at each watering, gradually release absorbed moisture to roots in contact with the pot's surface. This is especially beneficial for young seedlings. Clay pots also are more resistant to the growth of mold and fungi than are wood baskets, and they can easily be kept clean by an occasional blast of water into the pot. This practice also helps to prevent stem and root rot.

Small clay pots have long been used for very small seedlings after their removal from community pots. I was introduced to the use of clay pots for large seedlings and adult plants in 1992 by an innovative Thai grower, Pricha Tuvapalangkul, who was experimenting with using locally made, perforated clay pots as a substitute for teak baskets and was having excellent results. I have begun to follow his lead, also on an experimental basis, and have been pleased with the results. This method of growing should appeal to those who disapprove of using teak or cedar baskets for ecological reasons.

Coarse bark chips, tree-fern chunks and fibers, or hardwood charcoal may be used in clay pots. Coarse, shredded tree-fern fibers retain moisture longer than do chunks. It is best to combine the fibers with chunks of lava rock and chips of redwood bark or charcoal, for greater aeration.

Long roots cannot be intertwined in a clay pot the way they can with slatted wooden baskets; allow the roots to overhang the pot and dangle freely. Prune only in spring or early summer, if you feel you must prune.

## Plastic Pots

It is difficult to grow young vandaceous orchids with no upper aerial roots in plastic pots without some moisture-retaining medium. Without any medium, standard plastic pots drain almost instantaneously, are nonabsorbent, and restrict air movement around the base of the plant. To improve the flow of air around roots inside standard plastic pots, use an electric soldering iron, with the largest available head, to make a number of ventilation holes around the sides, and some extra ones on the bottom. When burning the holes, do it outdoors, because the fumes from many plastics are toxic. Air movement may have to be more powerful than when only slatted baskets are used, in order to compensate for the greater barrier effect of the pots. A better solution is to use plastic net pots, if you can obtain them in a sufficiently large size—their open-mesh construction allows excellent air circulation through the pots.

With plastic pots of any sort, use a medium that permits exceptionally good aeration and is not too retentive of moisture. Very coarse fir bark, cubes of coarse tree fern, or chunks of hardwood charcoal meet those tests. Some kinds of lava rock or expanded clay also serve, although, on an experimental basis, I have had little success with them as the sole medium, not only in my New York greenhouse but also outdoors in the mild, humid climate of the Canary Islands, off the coast of western Africa. Shredded tree-fern fibers, which retain moisture longer than some other media, work well in hot, arid climates.

## Moving Up to Larger Containers

When seedlings are ready to be taken from a community pot, they should be placed individually in small, shallow, slitted or perforated clay pots, or in small wood baskets. If you use a basket, tie the roots loosely to the bottom slats with a piece of cellophane twine or thin, plastic-coated wire (telephone wire is excellent), in order to keep the plants in place. Add potting medium if you wish (I don't). Anchoring plants in clay pots is not so easy; it requires a little ingenuity to tie them in place.

The time to move from a smaller to a larger wood basket, if it is to be done at all, is as soon as a plant begins to send its roots through the sides of its basket. If the roots are allowed to dangle freely, there is no reason to move to a larger basket as long as the original basket is in good condition. If it is not, it should be changed immediately.

If a larger basket is needed, and regardless of whether any medium has been used or not, remove the plant from its old basket and begin again with a larger new one. It is best not to merely insert the old basket into the new one, even if the old basket seems to be in good condition. Before removing the plant, soak the roots for at least five minutes in tepid water to make them more pliable and easier to detach from the old slats. Use a clean plastic plant label to help lift roots from the slats. Wash off the roots and remove any dead ones, using a sterile cutting instrument. While they are still pliable, work any long roots through the side or bottom slats of the new basket, rather than coiling them inside. The task is facilitated if you first remove one or more of the bottom slats from the new basket. They can be pried loose with a screwdriver or chisel, or you can bend the bottom wires to an upright position, lift off the two bottom runners, pry off the bottom slats, and put the runners back into their original position. Either way, the opening at the bottom of the basket will have been enlarged. Long roots can then be inserted through the opening. Restore the bottom slats to their original position by sliding them back into place over the runners. New roots will soon anchor them.

If the old basket is in excellent condition and you have not been using any potting medium, you may put the old basket into a new, larger one, but bear in mind that once an old basket is inside a new one, it becomes exceedingly difficult to check on its condition, and it becomes absolutely impossible to do so if the slats are surrounded by potting medium. That is why I counsel against putting the original basket inside the new one. But if you do that nonetheless, make sure that the new basket is sufficiently large to permit about 4 cm (1.5 in) of space around all of the sides of the old basket, for ventilation and ample room for new root growth. Octagonal baskets are an excellent choice for the second (or first) basket, because they allow more space and increased surface area to which roots can attach themselves. Always remove and discard the bottom slats (not the runners) of the smaller basket. Removing the slats also allows better ventilation and room for new roots. The underside of those slats would be the first to deteriorate and the hardest to examine.

When moving from a smaller to a larger clay pot (or to any other larger container), there is no need to first remove the plant. After soaking the pot thoroughly to soften the clinging roots and make them more flexible, crack the pot into several pieces with a hammer, exercising care to minimize damage to roots, and place the plant with any attached or unattached shards into the new container.

Never insert a smaller plastic pot into a larger one when the plant outgrows its pot. Soak the roots thoroughly, and lift them from the surface of the plastic, using a clean plastic label. Then move the plant to the larger pot. Wash the roots and cut off any dead ones, then fill the pot with fresh medium.

## GROWING SMALL SEEDLINGS

Sooner or later, most *Vanda* and *Ascocenda* growers buy a flask or community pot of promising seedlings. How to get such infants off to a good start is not something you necessarily learn from experience with more mature plants.

When you remove seedlings from the flask in which they were first grown, rinse them thoroughly with lukewarm water. Ordinarily, it is not necessary to add any disinfectant to the rinse, because the plantlets are sterile. (If there is any sign of mold, however, use a weak solution of Physan or its equivalent, or of Banrot or Natriphene.) Rinse two or three times in a bowl, then let them soak for about 10 minutes in a very mild solution of a well-balanced fertilizer (3 ml of fertilizer per 10 liters [0.25 tsp per gal] of water). Most of the nitrogen in the fertilizer should come from a nitrate source. It also is beneficial to add a like quantity of household sugar to the fertilizer.

Little seedlings from flask are extremely sensitive and vulnerable during their first 6–12 months out of flask. They insist on "togetherness," perhaps in order to preserve humidity around their roots. Do not be too quick to put them in individual containers. First plant them in a community pot.

A variety of growing media are suitable for use in community pots. A good conventional one is a mixture of coarse tree fern fiber, charcoal chips, redwood bark, and sponge rock or lava rock. Whatever material is used, it must maintain good aeration and not compact. An unconventional but satisfactory medium is small chunks and pieces of shredded polyurethane sponge, of the sort used in pillows and upholstery.

The "secret" to good growth is not so much the medium as an appropriate combination of watering, temperature, and light. In fact, I have had greatest success by arranging the plantlets loosely in a large plastic net basket, without any medium. Keep the plants in the community pots bathed with humid air but not continuously wet. If the thin roots dry out completely, the plants will go into shock and lie dormant for weeks or even months. On the other hand, they will rot very quickly if they remain wet for an extended period. You must find a way to provide high humidity combined with very gentle ventilation. Small fans can help.

Warmth and soft light also are required. Avoid large fluctuations in ambient temperature; a difference of 5°C (10°F) between night and daytime temperatures is ample, and a minimum temperature of 21°C (70°F) is ideal. The easiest way is to grow the seedlings under lights at first, indoors, rather than placing them in the greenhouse or outdoors immediately. In that way, the factors of light, temperature, and moisture can be better controlled.

When a community pot is purchased, the plantlets may be shipped in the pot or out of the pot. In either case, if the plants are so small that they should remain in a community pot, treat them like seedlings removed from flasks, and repot them in new medium (if you wish to use a medium). Usually, a few of the seedlings are much larger than the others and can be placed in individual pots; if so, that should be done. It is better for them, and it avoids their crowding the other, smaller seedlings in the community pot. Or group the largest seedlings together in a separate community pot.

When plants grown indoors under artificial lighting begin to outgrow their community pot and are almost ready for individual baskets or pots, place the community pot in a sheltered place in the greenhouse (or lathhouse), in order to harden the plants before translating them. In colder climates, do this only in late spring or early summer, and never in the late fall or winter; otherwise, the transition would subject the seedlings to too much stress. On sunny days, they may need to be misted or watered several times daily until thicker roots are formed, unless the surrounding atmosphere already is very humid.

Give small seedlings a very dilute fertilizer (about 3 ml per 10 liters [0.25 tsp per gal] of water) several times each week, until they are large enough to be treated the same as the more mature plants in the greenhouse or lathhouse.

## TOPPING

If the lower part of the stalk of a plant has no leaves, or if the plant has become too tall, and provided there are good roots on the upper portion of the stalk, you may sever it at an appropriate point and put the upper section of the plant in a new basket or pot. Keep the lower section, too, if you like the plant; it probably will send out new shoots from the base.

The first step is to remove all dead vegetative matter from the stalk. New roots are much more likely to break out from clean stalks than from those covered with dead tissue from the basal remains of old leaves. Dead tissue harbors spores of fungi and bacteria. Spores require a moist surface in order to germinate, and surfaces under dead tissue remain moist for a surprisingly long time, in comparison with live tissue exposed to air and sunlight. A strong blast of water will peel off the dead matter or at least loosen it enough so that it can easily be pulled off by hand. Always pull any clinging basal leaf tissue upwards or sideways; pulling it downwards often will tear off or expose some of the live tissue of the stalk, leaving a wound. If you do create a wound, seal it. A thin paste made of water mixed with equal parts of hydrated (mason's) lime and Banrot is good, but commercial horticultural sealants are available, and pastes can be made with other broad-based fungicides besides Banrot.

Make the cut about 1.5 cm (0.5 in) below a strong root and where there are at least two more well-developed roots there or higher up. There is no advantage in making the cut any lower, and it may be risky to cut much closer to the nearest root; on occasion, however, that is unavoidable. Bonsai shears are excellent for making the cut. Sterilize the cutting surfaces by flaming with a torch before each use.

Carefully examine the cross-sectional cut for signs of disease in the tissue of the stalk. Vandaceous orchids are susceptible to root rot and fusarium wilt. The pathogens enter the plant through dead roots, or through the hollow stubs of past inflorescences, and work their way upward through the vascular tissue. Infected areas are purple or brown in color. They resemble a discoloration more than an infection. On the cross-sectional cut, the visible signs may be no larger than the size of a small pinhead in the very center, or the discoloration may be a larger, irregularly shaped area on the perimeter. There will be no visible infected spots or areas on the exterior surface. If you find evidence of infection, keep cutting off higher sections of the stalk until there is no trace of the discoloration on the cross section. To be safe, soak the cut end of the stalk in a solution of fungicide, such as Banrot or Natriphene, for approximately 15 minutes; seal the cut once it has dried, and hope for the best.

Place the upper portion of the stalk in a new basket, but do not let the base of the stalk rest on a bottom slat. Position it between two of the slats, and keep the base of the stalk fully exposed to air. Let the lowest roots support the plant. Otherwise, the cut end of the stalk may absorb dampness from the adjoining slat as the sealant wears off, and bacteria or fungi may enter. Slatted baskets (without any medium) are better at preventing such problems than are pots containing a potting medium, no matter how coarse the material may be. If a clay pot is used (with or without potting medium), position the base of the stalk over or through the hole in the center of the pot, so that it will be exposed to air.

If the plant just topped is a prized one, retain the remaining lower half, provided it has live roots and no stem rot. One or more shoots usually will sprout from the stump, either near its top or near its base. Basal shoots develop roots much faster and grow more vigorously than do apical shoots. If you keep the lower half, seal the top of the stump and cut off any dead roots; seal those cuts also. You may remove the old basket at the same time and place the stump in a new one, or you may leave it in the old basket until you see signs of shoots emerging. Eric Christenson has found that it is best to let new shoots develop before repotting, because new root growth occurs after new shoots have been initiated. (It always is best to repot during periods of active root growth.) Clean out all detritus and sooty mold from the old basket, if it is kept in use; a strong jet of water is the most effective way, especially if a disinfectant has been added to the water.

## PESTS AND DISEASES

Vandas and ascocendas and their relatives are more trouble-free than most other orchids, but that does not mean that they are completely free of pests and diseases. Like orchids, fungi and bacteria thrive in warm, moist places; keeping your plants healthy requires constant vigilance and a program of preventive maintenance. Good sanitation and good cultural practices are essential for raising healthy orchids of any type. Regular monitoring of plants and containers will help you catch problems before they get out of hand.

A regular spraying program with a general purpose disinfectant will help prevent an accumulation of pathogens. Periodic treatments with a broad-spectrum fungicide and occasional drenchings with lime water also discourage fungi. Various chemicals are effective as insecticides. Frustratingly, the recommendations and permitted applications of specific chemicals change frequently, as new chemicals appear on the market and old ones are removed, sometimes for safety reasons. The "Question Box" section of the monthly *Bulletin* of the American Orchid Society is an excellent source of current information on cultural problems. The best general publication is *Orchid Pests and Diseases,* issued by the American Orchid Society (1995).

It is best to keep the populations of harmful fungi and other pathogens under control, rather than let them build up and then try to knock them out with one heavy blow. Use any chemical only when there is a reason for so doing. The specifications in this section for frequency of use are the maximum applications. The prophylactic treatments should be less frequent if there are no signs of any trouble whatsoever.

Once in a while, even with all your efforts at prevention, one or more of your plants will be attacked by pests or disease. If you are paying close attention to the health of your plants, you can institute a treatment program before the problem gets out of control. If, despite all efforts, a plant succumbs, learn from the experience. Think of all the new and potentially better hybrids that have been developed—plants you have wanted to own but could not accommodate because of a lack of space. Enjoy the prospect of upgrading your collection.

## Safety Concerns

First and foremost, protect yourself. Use protective clothing, goggles, and a mask designed to block organic vapors. Wash or bathe thoroughly after each use of toxic chemicals.

Never apply any chemicals under conditions of high heat and strong light; phytotoxicity may result. Phytotoxicity more often is caused by improper application than by any inherent incompatibility of the chemical and the plants affected. Apply chemicals early in the morning, before the temperature is high and the sun capable of scorching. Cloudy days are perhaps the best.

Carefully read and follow the instructions and cautionary statements on the label of all chemicals. Always be on the lookout for symptoms of damage or sickness that may be attributable to adverse reaction to chemicals you have been using. Remember that signs of toxicity do not always appear immediately; there may be a considerable delay. If you are suspicious, seek help—from growers, from the manufacturer, or from state agricultural services. Do not be surprised if you get conflicting answers in some instances; the problems can be complex and individual situations may be unique. For example, there may be an interaction between some of the chemicals you have been using, or your plants may have been stressed by climatic conditions and became especially susceptible to a chemical that, in other circumstances, would have caused no harm.

## The Importance of Sanitation

Provide a good environment and sanitary conditions for your plants. Pests and disease thrive in unsanitary conditions and a stale atmosphere. They strike weak plants more often and more forcefully than they do vigorously healthy plants. Chronic problems with pests and diseases usually indicate poor culture, and maintenance of sanitary conditions is an essential element of good culture.

Keep your plants and growing area clean. Remove all decaying vegetative matter in the leaf axils, along the stalks, in the baskets or pots, and on the floor. If decaying matter and other vegetative debris are not removed, growth of fungi and bacteria is a certainty under conditions of high humidity. Such debris also makes it difficult for disinfectants to work effectively. Blasts of water from a garden hose nozzle are a good way to loosen the remains of leaf tissue that cling to the stalk long after the leaves have dropped.

Detritus in leaf axils also provides a fertile medium for the seeds of oxalis, a common greenhouse weed that can damage vandas and ascocendas. When oxalis seedlings sprout in the leaf axils, which they will, their long, thin roots penetrate the base of the adjacent leaf. It will turn yellow, first on one side of the center rib, and eventually the entire leaf will die. The best prevention is to not allow a single plant of oxalis to exist in the growing area.

Stubs of past inflorescences are a frequent pathway for infection. Seal the stubs as soon as the flower stalks have been cut or broken off. Decaying stubs often go unnoticed, especially if the plants are hung high, because they are hidden in the axils of the leaves. The same is true of dead roots inside the baskets or pots. Whenever a dead root is spotted, cut it off with sterile shears and seal the cut. Dispose of fallen leaves and flowers as soon as you see them. Replace wooden baskets at the first sign of any decay. Inspect the interiors of your baskets and pots periodically. Decay usually begins on the inside surface of the bottom slats of baskets, where it is difficult to detect because of the presence of roots. The old maxim that "cleanliness is next to godliness" may or may not be true in general, but it certainly is relevant for consistently successful orchid growing.

## Diseases and Pests

### Sooty Mold and Algae

Sooty mold is a furry black substance that can cover the basal parts of roots and the slats of wooden baskets. It is unsightly but does not cause disease directly. It does, however, maintain moist conditions around the roots it embraces, and that may promote development of harmful fungi and bacteria. It also makes wood baskets rot faster, shortening their useful life.

Algae also does no harm to plants directly. Like sooty mold, it thrives under bright, warm, moist conditions, and is unsightly. It also makes walkways slippery, and when it forms on greenhouse coverings, it diminishes the passage of light. It should be removed as soon as it starts to build up.

The presence of any appreciable amount of sooty mold indicates that an adjustment in cultural conditions is needed. Poor air circulation and insufficiently long drying cycles are the usual causes. Peel off heavy accumulations of sooty mold on basket slats, using a plastic plant label, or blast it off with a jet of water from a garden nozzle. A drenching with a disinfectant and surfactant, such as Physan, R.D. 20, or Green-Shield, will kill the mold and help prevent its recurrence while you are adjusting the cultural conditions. These products also are effective in controlling algae. Use the manufacturer's recommended dosage. Generally that is 12–20 ml per 10 liters (1–1.5 tsp per gal) of water. A jet of water also helps remove algae from greenhouse coverings, but supplementary scrubbing with a brush usually is needed if the accumulation is heavy; follow with a rinsing, preferably with a disinfectant added to the water.

Physan, R.D. 20, and Green-Shield can be purchased from horticultural suppliers (often a supplier will carry only one of the brands). These three quaternary ammonium compounds are liquid surfactants, which reputedly work by disrupting the cell membranes of algae and fungi. They also dissolve some of the cuticle of the leaves (the noncellular layer of waxy substance that coats the surface of the leaves), thereby loosening the attached algae, fungi, and any mineral deposits and making it easier to flush them off. Because these products are not metabolic poisons, they are relatively safe for humans to use; they are standard hospital disinfectants. Pathogens do not develop resistance to them, because of the nature of their chemical action.

With both sooty mold and algae, prevention is easier than correction. A regular program of spraying with a general purpose disinfectant will prevent or substantially delay accumulations of sooty mold and algae. When spraying, wet all the surfaces in the greenhouse, including walkways, walls, foundations, and plants. Once a month should be sufficient, unless your cultural conditions need adjustment or the weather is continuously damp and dark. In the latter situation, twice monthly may be necessary. Use a fairly coarse nozzle with considerable pressure behind it. These are contact disinfectants with little or no residual action, so the spray must contact all surfaces of plants, baskets, and pots. As a safeguard against any chemical burning of tender root or leaf tips, I mist the plants lightly about 15–30 minutes before applying the disinfectant, and I always do the spraying early in the morning, before the temperature has risen very high. By taking these precautions, I have never had any problem with burning tender growth or with phytotoxicity.

Use a broad-spectrum fungicide as a preventive measure about once every six weeks. A separate drenching, in between, with a lime-water solution of 40 ml of hydrated (mason's) lime per 10 liters (1 tbp per gal) of water is beneficial in discouraging growth of algae and fungi. Drench the plants with plain water first, to minimize absorption of the lime by the roots. Most of the lime in the solution does not dissolve, but enough dissolves or stays in suspension if the solution is stirred frequently. The slight residue of lime left in the leaf axils helps to prevent renewed growth of algae and mold in the axils, so drench heavily enough to flush the axils with the solution.

### *Phyllostica*

A malady increasingly being encountered is an unsightly fungus-induced leaf spot disease. In various publications, it has been referred to under several names: *Phyllosticta, Phyllostictina,* and *Guignardia**. In the late 1970s, it was discovered to be prevalent in the Rokeratchaburi Province in Thailand, and it then spread throughout that country and appeared in the United States shortly afterwards. It very probably is to be found wherever vandas and ascocendas have been imported from Thailand. It is characterized by very dark purple or black streaks that run parallel to the vein of the leaf. The streaks gradually widen and coalesce. On the top surface of the leaf, they can easily be distinguished from other leaf spot diseases by their raised, rough texture.

The disease spreads to other plants, especially under cool, damp conditions. While the infection is not vascular (Kulchawee Kamjaipai 1990), a Thai plant pathologist who specializes in orchid diseases, states, "affected plants always die prematurely." Perhaps the reason is that the infected leaves drop off and the plant gradually is weakened by their loss.

The best approach is never to accept a diseased plant. A vendor who minimizes the disease is doing you a grave and self-serving disservice. If you do encounter the disease in your collection, immediately cut off every infected leaf with a sterile blade. A weekly or fortnightly program of use of a horticultural disinfectant has a reasonable probability of eventually preventing recurrences of the disease, provided you have promptly removed every visibly infected leaf and have regularly sprayed all surfaces of every leaf of every plant in your collection.

### Fusarium Wilt

Fusarium wilt pathogens (*Fusarium oxysporum*) enter vandas and ascocendas through dead roots or through stubs of inflorescences and work their way upward through vascular tissue, particularly the xylem. The disease cannot be identified

---

*The AOS publication *Orchid Pests and Diseases* (1995) uses *Guignardia*.

externally with any certainty, because no external discoloration occurs and the symptoms mimic those of environmental origin. The leaves of an infected plant wither, and there is little or no growth of the roots. A diagnostic symptom sometimes occurs: the plant sends out side shoots high up on the stalk, and the uppermost leaves look rather strange; they are undulated or furrowed along their length. (Note, however, that furrowed leaves may be genetic in origin.)

To prevent conditions favorable to the development and spread of *Fusarium oxysporum* spores, use only nitrate nitrogen rather than blends containing ammoniacal or urea forms of nitrogen, and maintain water pH at 7.0 or somewhat higher. Research suggests that "an ammonium source of nitrogen (ammonium nitrates, ammonium sulfate) increases the development of Fusarium wilt," as does a pH below 7.0 (Engelhard and Woltz 1976). My experience shows that drenching with lime water helps to prevent the development of the spores in the leaf axils and around the roots. Use hydrated lime in a 40 ml per 10 liters (1 tbp per gal) solution. Stir the solution frequently while applying. Saturate the roots with plain water first, to minimize absorption of the lime water into the plants. Apply the treatment once every two weeks or so, depending on how dull and damp the weather has been. Dark, damp conditions, and absence of sufficiently long and frequent drying cycles, are conducive to spore growth; so is insufficient air movement.

If a plant exhibits any of the described symptoms, suspect vascular infection within the stalk, especially if there are no obvious cultural explanations. Cut off a short piece of the plant stalk at its base, and examine it for telltale spots of purple or brown discoloration. Bear in mind that the discolored area may not be any larger than the head of a pin, especially if it is in the center of the cross section. If the plant is infected, cut off small pieces of the stalk as described in the Topping section of this chapter, until the cut end is completely free of discoloration. Soak the plant in a solution of a broad-spectrum fungicide, such as Banrot or Natriphene. Seal the cut after it has dried. If you cannot outrun the discoloration while there still are roots above, hang the plant upside down, and hope that the fungicide treatment will work as a cure. The odds in your favor are small but greater than zero.

## Viral Diseases

Vandas and ascocendas, like other orchids, can be infected by plant viruses. Infection almost always is spread from one plant to another by contaminated cutting instruments, although there are reports that some insects, such as aphids and thrips, can transmit some kinds of viruses when the insects penetrate plant tissue. There are no cures for virus, only preventive measures. Use only sterile tools. Keep your growing area clean and as free from pests as possible. Ideally, wash your hands before and after handling each plant if you have any reason to think that some of your plants may be infected.

Viral diseases are difficult to diagnose with any certainty without laboratory testing. The symptoms vary, and some resemble those caused by other pathogens or by environmental conditions. If in doubt, and certainly before discarding a valuable plant, send leaves to a testing laboratory that is expert in the matter; several advertise in the *Bulletin* of the American Orchid Society. Isolate the plant until you know the results. Discard it if the results are positive.

### Insects

Keep insects under control. Scale, mites and mealybugs are common pests in greenhouse environments. They cause damage by sucking juices from the plant. In some areas, thrips and aphids also are a problem. Some insects spread virus, according to some reports.

Horticultural oil is a simple and effective way to control nearly all insects commonly found in greenhouse. One such product, widely used by orchid growers, is Safer's SunSpray Ultra-Fine Spray Oil. The oil kills mainly by smothering the insects, rather than by poisoning them, so it will not kill all stages of development; two or three applications, a week or 10 days apart, are needed for satisfactory control. If insect infestation is limited to a few plants, many growers spray only those, using a half-liter or one-pint bottle sprayer.

Follow the dosage recommended by the manufacturer. Oil in a water solution rises to the top quickly, so shake or stir the solution frequently while using. Apply the oil early in the day, with a light but uniform coverage. Be sure to cover all leaf and plant surfaces, but there is no need to drench the plants unless there are mealybugs in the potting medium (if any is used). A warm, but not hot, cloudy day is best; warmth speeds evaporation, which helps to minimize any risk of phytotoxicity from the oil, but high temperature can cause stress in the leaves, because the oil reduces the rate of transpiration of the plants, by covering the stomata of the leaves. It is a good idea to rinse the plants thoroughly with plain water the following day, in order to wash off any residue of the oil and restore normal transpiration; that, too, minimizes risk of phytotoxicity. The rinse will be more effective if a surfactant (e.g., Physan) has been added to the rinse water. The oil already will have smothered any insects it contacted, so as much of it as possible should be removed. After a week or 10 days, give the plants a second application of the oil. Occasionally, a third will be needed, after a similar interval, particularly in situations in which it is difficult to get the spray on all the undersurfaces of the leaves. That is almost always the case where there is overcrowding; in that situation, adding a small dosage of an insecticide such as Malathion, Isotox, or Orthene greatly improves the effectiveness of the oil, but do so only when the oil alone is inadequate.

By following these recommendations, you can largely avoid use of metabolic poisons as insecticides. Horticultural oil, unlike most other insecticides, is not a metabolic poison, and is much less toxic than such products. Even so, take the same personal precautions with horticultural oil that you would when using any other pesticide.

## Other Pests

Ants can be a major problem in many areas at certain times of the year. While they do no direct harm to the orchids, they transport young scale, aphids, and mealybugs from one plant to another. It is hard to control these three insects without controlling the ant population. The best way is with ant bait, not with insecticides. The ants carry the poisoned bait from the containers back to their nests, where it is consumed by other members of the colony, too. The baited containers can be purchased at most hardware or garden supply stores. Merely place them in strategic locations. Cover the containers in some way so that they do not get wet when you are watering. These devices are simple and effective. Put them in place at the first signs of ants in the spring, and do not skimp on the number.

Insects, fungi, molds and bacteria are not the only threats to orchids. Mice, chipmunks and even squirrels can raise havoc by eating parts of the buds and flowers or by knocking over plants—always our most prized ones, of course. Poison bait or traps often are necessary, especially in the fall of the year when rodents seek warmth and food inside the greenhouse.

## TREATMENT OF STRESSED PLANTS

Any plant that shows signs of serious stress, whether from Fusarium wilt or any other cause, will benefit from several weeks or a few months of the upside-down treatment. Bareroot the plant, wash the roots, and cut off any dead ones. Soak the plant in a glucose solution unless the problem appears to be of fungal origin; in the latter case, soak the plant in a solution of Banrot or Natriphene. Then suspend the plant by a wire or string, upside down. Place it in a shady, humid place, with mild, but not desiccating, air movement.

Hanging the plant in this way prevents water from lingering in the leaf axils, thus discouraging the growth of fungi and bacteria. The principal benefit, however, comes from a reversal of apical dominance. Hanging the plant upside down thwarts apical dominance by reversing the direction of the internal flow of hormones that normally inhibit root and lateral shoot growth in order to provide more energy to the top of the plant. As a consequence, energy is diverted from the production of new growth at the old top of the plant to production of new roots at what now is the uppermost section. Some growers like to place a plastic bag over the upside-down plant, to conserve humidity. Under my conditions, that method seems to increase the occurrence of rot.

Once new roots have emerged from the stalk and have reached a length of 3 cm (1 in) or more, put the plant in a basket and hang it right side up. Gradually increase the amount of light to which the plant is exposed. If you delay uprighting the plant too long, the stalk of the plant will curve back upwards as new leaves form, and you will end up with a U-shaped stalk.

# Appendices

# A

# A Guide to Orchid Classification and Nomenclature

Almost every field of endeavor has some system or systems of terminology and classification the purpose of which is to facilitate clarity and precision of communication. Botany is no exception. Horticulturists rely heavily on the nomenclature and classifications established by botanists; however, horticultural practices often differ from, and lag far behind, the taxonomy of botanists.

For orchids, the standard horticultural nomenclature is that established by the *Handbook on Orchid Nomenclature and Registration,* prepared by the International Orchid Commission (1993). It is not, however, authoritative on matters of botanical taxonomy; only on matters of registration of hybrids. (For botanical nomenclature, see the *International Code of Botanical Nomenclature,* 1988, adopted by the Fourteenth International Botanical Congress.)

Botanists employ a hierarchical system for grouping plants and use an accompanying nomenclature that follows rather precise rules. The aim of the nomenclature, as stated by the International Orchid Commission in its *Handbook,* "is to provide every kind of plant with an internationally agreed name that applies only to that particular type of plant."

For example, once one is acquainted with the system, one can tell by a glance at the name of an orchid in what genus it belongs, and whether the name refers to a species or a hybrid. One also can tell whether the name refers to a single individual plant in a group or to the group as a whole.

For wild orchids, the system of hierarchial classification is based on shared characteristics of morphology. The term *taxon* (plural, *taxa*) is a general one that can be applied to any group of any rank in the hierarchy. It is, in effect, the unit of rank.

## THE HIERARCHY OF TAXA AND THEIR NAMES

At the highest and broadest taxonomic level, all orchids are placed in the taxonomic rank of "order." The name of that order is Orchidales. Descending from order, are these ranks: family, subfamily, tribe, subtribe, genus, subgenus, species, subspecies, variety, form, and cultivar—to name the principal ones.

Names of taxa above the rank of genus generally are written in roman print (i.e., are not italicized), and the taxon always begins with a capital letter. The taxon is a uninomial, a single word. To illustrate, the genus *Vanda* belongs in the Sarcanthinae subtribe of the Vandeae tribe; Sarcanthinae and Vandeae are the taxa ranking immediately above the level of genus.

Under the levels of genus and subgenus, orchid names always consist of a minimum of two *terms*. Some kinds of terms may consist of only a single word, while some others may contain more than one word. Some are italicized, some have initial capital letters, while some others are printed in roman type or do not begin with a capital. Differences in these details are part of the code and must be observed if confusion is to be avoided.

### Genus and Species

The first term in the name of an orchid indicates the *genus* in which the plant belongs; it is not customary to state the taxa above the rank of genus. A genus is described in the *Handbook of Orchid Nomenclature and Registration* (International Orchid Commission, 1993, 139) as being

> a subdivision of a family, consisting of one or more species which show similar
> significant characteristics and appear to have a common ancestry.

The expression "common ancestry" means far back in time, not a few decades or centuries.

The name of a genus, usually called the *generic name,* always consists of a single word. The word is written in Latin form, begins with a capital letter, and is italicized. These rules apply both to natural genera, such as *Cattleya, Phalaenopsis,* and *Vanda,* and to man-made or "artificial" hybrid genera, such as *Ascocenda* (a genus formed by combining *Ascocentrum* and *Vanda,* both of which are natural genera).

The generic name should not be pluralized by adding an "s" at the end of the term—the same form serves as the singular and the plural. In horticultural writings, an "s" sometimes is added, but in that case the generic term no longer is italicized and it begins with a lower-case letter, as in "ascocendas."

If a plant is a *species,* the second term in the orchid name is the *species epithet,* more commonly called the *specific epithet*. The word "species" is both singular and plural. (The word "specie" refers to money.) A species is defined in the *Handbook* (141) as

> a group of plants (or animals) showing intergradation among its individual members and having in common one or more significant characteristics which definitely separate it from any other group.

The specific epithet always is a single word. It is written in Latin form, and begins with a small letter even when the species is named after an individual—*sanderiana* or *bensonii,* to take two *Vanda* species named after individuals. These rules apply to the epithets of subspecies as well. It should be noted that names derived from persons should be pronounced like the person's name. For example, *Vanda bensonii* has the accent on the first syllable of the epithet, not the second.


## The Grex Epithet

If a plant is an artificial (man-made) hybrid, however, the second term in the name is called the *grex epithet*. This term may also be called the *collective epithet,* which is the expression used in botanical literature for plants other than orchids, and is valid for natural hybrids as well. The grex epithet of a man-made hybrid never is italicized, and it may consist of more than one word, each of which begins with a capital letter.

Thus, at a glance, one can tell that the name *Vanda coerulea,* for example, refers to a species in the genus *Vanda*, while *Phalaenopsis* Stewart's Pride is a man-made hybrid within the genus *Phalaenopsis* (i.e., an intrageneric hybrid).

The name of a species or hybrid, therefore, is a combination of two terms: in the case of a species, the generic name plus the specific epithet, and, in the case of an artificial hybrid, the generic name plus the grex epithet. For artificial hybrids, the full name may be referred to as the *name of the cross* (i.e., of a pair of parent plants), instead of the grex name.

All plants with the same pair of parental names share the same grex or collective name. The *individual* parent-plants may be different, but the progeny will share the same grex name as any other plants whose parents bear that same pair of names. From the standpoint of naming, it makes no difference which plant served as the seed-bearing (female) parent and which served as the pollen (male) parent. For example, *Vanda* Rothschildiana may be a cross of either *Vanda coerulea* and *Vanda sanderiana* or the opposite order.

When a new cross is registered, however, the seed-bearing parent's name is listed first, but if the cross is remade with the parentage roles reversed, the progeny still will retain the grex name (*Vanda* Rothschildiana, in the example cited). A cross of one *Vanda* Rothschildiana plant with another would produce more *Vanda* Rothschildiana; the offspring may not be given a different grex name.

## Form, Variety, and Subspecies

Three ranks below that of species frequently are encountered and often are misused. These are *form* (abbreviated "f."), *variety* (abbreviated "var."), and *subspecies* (abbreviated "ssp.").

A *form* is a sporadically occurring mutation occasionally found among the wider population of the species. Examples are the *alba* and *aurea* color forms. These mutations often are prized horticultural plants, but they have little impact on the genetic evolution of the species. Another example is the peloric flower form of some *Phalaenopsis* and *Cattleya* species and hybrids. In the genus *Vanda*, *V. sanderiana* var. *alba* is a name commonly seen, but it is not correct; the correct name is *V. sanderiana* f. *alba,* or "the *alba* form of *V. sanderiana.*"

A *variety* of a species is a definable variant that constitutes a fraction of the general population and is a stable part of the genetic structure of the species, unlike the ephemeral nature of forms of the species. *Vanda suavis* var. *boxallii* is an example. A variety does not have its own geographic range, unlike the case of subspecies.

A *subspecies* is a geographic race. Not only does it have definable morphological differences from the species type but it also has geographic boundaries that divide the members of the subspecies from the other members of the total population. For example, evolution on an isolated island might lead to differences from the members of the species that evolved elsewhere. A specific example would be *Ascocentrum aurantiacum* (Christenson, 1992). This species originally was described from Sulawesi (Celebes), but the flowers of the plants that developed in the Philippines are smaller in size and are treated as *A. aurantiacum* ssp. *philippinensis* Christ. According to Christenson (Christenson, 1994):

> Subspecies are generally viewed as genetically compatible parts comprising a single species. In theory, should the (geographical) barriers disappear, the subspecies would meld and the morphological distinctions disappear. On the other hand, if the isolation persists (over an evolutionary significant period), the subspecies may evolve into distinct species due to the incompatibility formed by divergent morphological and genetic changes.

## Cultivar or Clonal Epithets

Descending still farther down the hierarchy of taxa we come to *cultivar* or *clone;* both words are used. The cultivar or clonal epithet identifies a particular individual plant (and divisions and artificially produced clones thereof, because genetically the latter, barring mutations, normally will be identical). A cultivar epithet may consist of more than one word. Each word is printed in roman type and begins with a capital letter. The epithet is enclosed in single quotation marks. An example would be *Vanda sanderiana* 'Riopelle's Treasure'. In this coined example, the entire combination would constitute the cultivar or clonal *name* of the plant.

Unlike the case where one plant of a grex is crossed with itself or with another member of the same grex, and transmits the original grex name to all the progeny of the mating, a selfing of a cultivar (or a cross of it with a division or mericlone thereof) does not transmit its clonal name. The reason is that the resulting offspring will not be genetically identical; in fact, they may demonstrate considerable variation from one specimen to another.

### Intergeneric Hybrids

The discussion to this point has concentrated on the nomenclature of species and of artificial crosses of orchids falling in the same genus (i.e., *intra*generic crosses), but this book deals with many genera formed by humans by combining two or more natural genera. All the orchids we discuss in this book belong in genera grouped in the subtribe Sarcanthinae (syn.: Aeridianae) of the Vandeae tribe—two taxa immediately above that of genus. The subtribe Sarcanthinae encompasses the natural genera *Aerides, Ascocentrum, Phalaenopsis* (including *Paraphalaenopsis*), *Renanthera, Rhynchostylis,* and *Vanda,* among others. It is a very large group, and many artificial intergeneric hybrids have been made within the subtribe.

An example of such an intergeneric hybrid is *Ascocenda.* It is a cross that contains genes of both *Ascocentrum* and *Vanda,* each of which is a separate genus within the Sarcanthinae subtribe. Note that, like any other generic name, it is uninomial, italicized, and written with an initial capital. The rules of nomenclature for man-made intergeneric hybrids are the same as for those that are intragenic.

There can be much greater variability of floral characteristics among intergeneric hybrids than within the component individual genera, simply because a larger number of species is available for inclusion. That is why the potential of an intergeneric grex generally cannot be anticipated without some genealogical and comparative research—there is a multitude of possible combinations of inputs. One has to take into account not only the number and kind of species involved but also the frequency of inclusion of each one, and the degree of remoteness in the ancestry.

Some man-made intergeneric combinations are very complex. For example, *Darwinara* is the generic name for any combination (regardless of order or frequency) of plants from the four genera *Ascocentrum, Neofinetia, Rhynchostylis,* and *Vanda,* most of which, in turn, contain a number of individual species. The range of possible combinations of genes in such multigeneric mixtures is formidable. That is part of their allure.

## Identification of the Parents' Names

The names of a man-made hybrid's two parents often are stated on plant labels and in publications, generally with the multiplication sign "×" written in between them, and the whole expression is enclosed within parentheses. For example, *Ascocenda* Yip Sum Wah is a cross of the intrageneric hybrid *Vanda* Pukele and the species *Ascocentrum curvifolium*. Typically, this would be written *Ascocenda* Yip Sum Wah (*Vanda* Pukele × *Ascocentrum curvifolium*). *Vanda* Pukele would be the pod parent and *Ascocentrum curvifolium* would be the pollen parent of the cross, in this example.

Whenever both parents of a grex belong in the same genus, it is not necessary to repeat the name of the genus when giving the name of the parents, although it is clearer to do so (and I do so in this book). For example, it is acceptable to write *Vanda* Orglade's Rosy Dawn (Charungraks × coerulea). That is the practice followed in *Sander's List of Orchid Hybrids*.

## Natural Hybrids

The multiplication sign also is used in another context. Sometimes nature produces a natural *intra*generic hybrid. One such is a hybrid of *Vanda coerulea* and *Vanda bensonii*. As a hybrid, one might expect the collective epithet to be *V.* Charlesworthii. However, it is written as *Vanda* × *charlesworthii,* like the case of the collective epithet of a species (i.e., no initial capital and italicized). By putting a multiplication "×" in front of the second term and by writing it in the same form as a specific epithet, the signal is given that it is a natural, not a man-made, hybrid. It should be mentioned that there are few natural hybrids, and, in the Sarcanthinae subtribe, none, to my knowledge, are intergeneric—only intrageneric.

## AUTHORS' NAMES

In botanical, as contrasted with horticultural, literature, it is customary to follow the *initial* use of the name of a species or natural genus with the name of the person who first described and named the species or genus. That person is referred to as the "author." The author's name is printed in roman type and follows the name of the species or genus without any punctuation in between. An example would be *Aerides* Loureiro. In this instance, "Loureiro" does not stand for the grex epithet of an *Aerides* hybrid; instead it stands for João de Loureiro, who first described and published the description of the genus *Aerides* in acceptable botanical terms.

Often, the author's name is abbreviated (e.g., Loureiro is written as "Lour."). In nonscientific publications, it generally is omitted altogether. Sometimes more than one person's name is shown. That is done when someone subsequently has revised the original description.

The primary purpose of the practice of giving the author's name is not to spread the fame of the originator of the name of the species or genus; rather, it is to provide an accepted reference to which other botanists can turn whenever they wish to identify and properly classify a given specimen.

In this book. we give the author's name in abbreviated form at the beginning of the treatment of a species or genus. The abbreviated form also is followed in Christenson's list of the genus *Vanda* (Appendix B). In most instances, authors' full names can be found in Brummitt and Powell (1992).

# B

# Christenson's List
# of the Genus *Vanda*

**VANDA JONES ex R. BROWN**
A broadly defined *Vanda* including *Euanthe* and *Trudelia*

* indicates AOS "recognized"
‡ indicates served as a parent of a *Vanda* hybrid prior to year-end 1994.

## Species in *Vanda*

‡  *V. alpina* (Lindl.) Lindl.
    *V. arbuthnotianum* Kraenzl.
    *V. arcuata* J. J. Sm.
*‡ *V. bensonii* Batem.
    *V. bicolor* Griff.
    *V. × boumaniae* J. J. Sm. (*insignis × limbata*)—a natural hybrid
‡  *V. brunnea* Reichb. *f.*
    *V. celebica* Rolfe
    *V. chlorosantha* (Garay) E. A. Christ.
*‡ *V. coerulea* Griff. ex Lindl.
*‡ *V. coerulescens* Griff.
*‡ *V. concolor* Bl. ("Recognized" by the American Orchid Society as *V. furva*)
    *V. × confusa* Rolfe (*coerulescens × lilacina*)—a natural hybrid
    *V. crassiloba* Teijsm. & Binn. apud J. J. Smith ?
*‡ *V. cristata* Lindl.
*‡ *V. dearei* Reichb. *f.*
*‡ *V. denisoniana* Benson & Reichb. *f.*
    *V. devoogtii* J. J. Smith
    *V. flavobrunnea* Reichb. *f.* ?
‡  *V. foetida* J. J. Sm.
    *V. fuscoviridis* Lindl.

    *V. gibbsiae* Rolfe

    *V. griffithii* Lindl.

\*   *V. hastifera* Reichb. *f.*

‡   *V. helvola* Bl.

\*‡ *V. hindsii* Lindl.

\*‡ *V. insignis* Bl.

    *V. jainii* Chauhan

    *V. javieriae* Tiu ex Fessel & Lückel

\*‡ *V. lamellata* Lindl.

    *V. leucostele* Schltr.

‡   *V. lilacina* Teijsm. & Binn.

\*‡ *V. limbata* Bl.

    *V. lindeni* Reichb. *f.*

‡   *V. liouvillei* Finet

    *V. lombokensis* J. J. Sm.

\*‡ *V. luzonica* Loher ex Rolfe

\*‡ *V. merrillii* Ames & Quisumb.

    *V. petersiana* Schltr.

\*‡ *V. pumila* J. D. Hook.

    *V. punctata* Ridl.

\*‡ *V. roeblingiana* Rolfe

    *V. scandens* Holtt.

\*‡ *V. sanderiana* Reichb. *f.*

\*‡ *V. stangeana* Reichb. *f.*

    *V. subconcolor* Tang & Wang

‡   *V. sumatrana* Schltr.

\*‡ *V. tessellata* (Roxb.) W. J. Hook. ex G. Don

\*‡ *V. testacea* (Lindl.) Reichb. *f.*

    *V. thwaitesii* J. D. Hook.

\*‡ *V. tricolor* Lindl.

    *V. vipanii* Reichb. *f.*

## Species Synonymized in or Transferred from *Vanda*

*V. amesiana* Reichb. *f.* = *Holcoglossum amesiana* (Reichb. *f.*) E. A. Christ.

*V. amiensis* Masamune & Segawa = a misspelling of *V. yamiensis* Mas. & Seg. = *V. lamellata* Lindl.

*V. batemanni* Lindl. = *Vandopsis lissochiloides* (Gaud.) Pfitz.

*V. bicaudata* Thw. = *Diploprora championii* J. D. Hook.

*V. boxallii* (Reichb. *f.*) Reichb. *f.* = *V. lamellata* Lindl.

*V. cathcarti* Lindl. = *Esmeralda cathcartii* (Lindl.) Reichb. *f.*

*V. clarkei* (J. D. Hook.) N. E. Br. = *Esmeralda clarkei* (J. D. Hook.) Reichb. *f.*

*V. clitellaria* Reichb. *f.* = *V. lamellata* Lindl.

*V. congesta* Lindl. = *Acampe congesta* (Lindl.) Lindl.

*V. cumingii* Lodd. = *V. lamellata* Lindl.

*V. denevei* hort. ex Zurowetz = *Paraphalaenopsis denevei* (J. J. Sm.) Hawkes

*V. densiflora* Lindl. = *Rhynchostylis gigantea* (Lindl.) Ridl.

*V. doritoides* Guill. = *Ornithochilus delavayi* Finet

*V. esquirolei* Schltr. = *V. concolor* Bl.

*V. falcata* Beer = *Neofinetia falcata* (Thunb.) Hu

*V. fasciata* Gardn. ex Lindl. = *Acampe praemorsa* (Roxb.) Blatter & McCann

*V. fimbriata* Gardn. ex Thw. = *Gastrochilus acaulis* (Lindl.) Ktze.

*V. furva* Lindl. = *V. concolor* Bl.

*V. gigantea* Lindl. = *Vandopsis gigantea* (Lindl.) Pfitz.

*V. guangxiensis* Fowlie = *V. concolor* Bl.

*V. guibertii* Lindl. = *Staurochilus guibertii* (Lindl.) E. A. Christ.

*V. hainanensis* Rolfe = *Rhynchostylis gigantea* (Lindl.) Ridl.

*V. henryi* Schltr. = *V. denisoniana* Bens. & Reichb. *f.*

*V. hookeri* hort. = *Papilionanthe hookeriana* (Reichb. *f.*) Garay

*V. hookeriana* Reichb. *f.* = *Papilionanthe hookeriana* (Reichb. *f.*) Garay

*V. kimballiana* Reichb. *f.* = *Holcoglossum kimballianum* (Reichb. *f.*) Garay

*V. kwantungensis* Cheng & Tang = *V. fuscoviridis* Lindl.

*V. laotica* Guill. = *V. lilacina* Teijsm. & Binn.

*V. lindleyana* Griff. ex Lindl. & Paxt. = *Vandopsis gigantea* (Lindl.) Pfitz.

*V. lissochiloides* Lindl. = *Vandopsis lissochiloides* (Gaud.) Pfitz.

*V. longifolia* Lindl. = *Acampe rigida* (Buch.-Ham. ex J. E. Sm.) Hunt

*V. lowii* Lindl. = *Dimorphorchis lowii* (Lindl.) Rolfe

*V. masperoae* Guill. apud Gagnep. = *Papilionanthe pedunculata* (Kerr) Garay

*V. moorei* Rolfe = *(Holcoglossum kimballianum* × *V. coerulea).*

*V. muelleri* (Kraenzl.) Schltr. = *Vandopsis muellerei* (Kraenzl.) Schltr.

*V. multiflora* Lindl. = *Acampe rigida* (Buch.-Ham. ex J. E. Sm.) Hunt

*V. nasugbuana* Parsons = *V. lamellata* Lindl.

*V. obliqua* Wall. ex J. D. Hook. = *Gastrochilus obliquum* Lindl.

*V. paniculata* (Ker.-Gawl.) R. Br. = *Cleisostoma paniculatum* (Ker.-Gawl.) Garay

*V. parishii* Veitch & Reichb. *f.* = *Hygrochilus parishii* (Veitch & Reichb. *f.*) Engl. & Prantl

*V. parviflora* Lindl. = *V. testacea* (Lindl.) Reichb. *f.*

*V. peduncularis* Lindl. = *Cottonia peduncularis* (Lindl.) Reichb. *f.*

*V. pseudo-coerulescens* Guill. = *Rhynchostylis coelestis* Reichb. *f.*

*V. pulchella* Wight = *Gastrochilus acaulis* (Lindl.) Ktze.

*V. pusilla* Teijsm. & Binn. = *Trichoglottis pusilla* (Teijsm. & Binn.) Reichb. *f.*

*V. recurva* W. J. Hook. = *Cleisostoma rostratum* (Lindl.) Garay

*V. rostrata* Lodd. = *Cleisostoma rostratum* (Lindl.) Garay

*V. roxburghii* R. Br. = *V. tessellata* (Roxb.) W. J. Hook. ex G. Don

*V. rupestris* Hand.-Mazz. = *Holcoglossum rupestre* (Hand.-Mazz.) Garay

*V. saprophytica* Gagnep. = *Holcoglossum saprophyticum* (Gagnep.) E. A. Christ.

*V. saxatilis* J. J. Sm. = *V. lindeni* Reichb. *f.*

*V. scripta* Spreng. = *Grammatophyllum speciosum* Bl.

*V. simondii* Gagnep. = *Cleisostoma simondii* (Gagnep.) Seid.

*V. spathulata* (L.) Spreng. = *Taprobanea spathulata* (L.) E. A. Christ.

*V. stella* hort. = *V. concolor* Bl.

*V. storiei* Storie ex Reichb. *f.* = *Renanthera storiei* Reichb. *f.*

*V. striata* Reichb. *f.* = *V. cristata* Lindl.

*V. suaveolens* Bl. = *V. tricolor* Lindl.

*V. suavis* Lindl. = *V. tricolor* Lindl.

*V. suavis* F.v.Muell. = *V. hindsii* Lindl.

*V. subulifolia* Reichb. *f.* = *Holcoglossum subulifolium* (Reichb. *f.*) E. A. Christ.

*V. sulingii* Bl. = *Arachnis sulingii* (Bl.) Reichb. *f.*

*V. superba* Lind. = *V. lamellata* Lindl.

*V. taiwaniana* Ying = ?natural hybrid *(Luisia × Papilionanthe)*

*V. teres* Lindl. = *Papilionanthe teres* (Roxb.) Schtlr.

*V. teretifolia* Lindl. = *Cleisostoma simondii* (Gagnep.) Seid.

*V. tesselloides* Reichb. *f.* = *V. tessellata* (Roxb.) W. J. Hook. ex G. Don

*V. trichorhiza* W. J. Hook. = *Luisia trichorhiza* (W. J. Hook.) Bl.

*V. tricuspidata* J. J. Sm. = *Papilionanthe tricuspidata* (J. J. Sm.) Garay

*V. truncata* J. J. Sm. = *V. lindsii* Lindl.

*V. undulata* Lindl. = *Vandopsis undulata* (Lindl.) J. J. Sm.

*V. unicolor* Steud. = *V. lamellata* Lindl. and *V. tessellata* (Roxb.) W. J. Hook. ex G. Don

*V. vidalii* Boxall ex Naves = *V. lamellata* Lindl.

*V. viminea* Guill. = *Acampe rigida* (Buch.-Ham. ex J. E. Sm.) Hunt

*V. violacea* Lindl. = *Rhynchostylis gigantea* ssp. violacea (Lindl.) E. A. Christ.

*V. vitellina* Kraenzl. = *V. testacea* (Lindl.) Reichb. *f.*

*V. watsoni* Rolfe = *Holcoglossum subulifolium* (Reichb. *f.*) E. A. Christ.

*V. whiteana* Herbert & Blake = *V. hindsii* Lindl.

*V. wightiana* Lindl. ex Wight = *Acampe praemorsa* (Roxb.) Blatter & McCann

*V. yamiensis* Mas. & Segawa = *V. lamellata* Lindl.

## Names of Uncertain Placement

*V. cruenta* Lodd.

*Aer. flabellata* Rolfe ex Downie

*V. massaiana* hort. ex L. Linden

*V. pauciflora* Breda

# C

## Score Sheet of the American Orchid Society for Judging Flower Quality of Vandas

The same scale is used for ascocendas and most other members of the Sarcanthinae subtribe, except *Phalaenopsis*

|  | POINTS |
|---|---|
| Form of flower | 30 |
| General form (15) | |
| Sepals (7) | |
| Petals (5) | |
| Labellum (3) | |
| | |
| Color of flower | 30 |
| General color (15) | |
| Sepals (7) | |
| Petals (5) | |
| Labellum (3) | |
| | |
| Other characteristics | 40 |
| Size of flowers (10) | |
| Substance and texture (10) | |
| Habit and arrangement of inflorescence (10) | |
| Floriferousness (10) | |
| | |
| Total Points | 100 |

# D

## Registered Intergeneric Combinations with Strap-Leaf Vandas

Through year-end 1992 and excluding ascocendas

| *Vanda* | REGISTRATIONS |
|---|---|
| × Acampe = Vancampe | 2 |
| × Aeranthes = Vandaeranthes | 1 |
| × Aerides = Aeridovanda | 96 |
| × Aerides × Arachnis = Burkillara | 11 |
| × Aerides × Arachnis × Ascocentrum = Lewisara | 12 |
| × Aerides × Arachnis × Vandopsis = Pehara | 1 |
| × Aerides × Ascocentrum = Christieara | 61 |
| × Aerides × Ascocentrum × Neofinetia = Micholitzara* | 0 |
| × Aerides × Ascocentrum × Phalaenopsis = Isaoara | 1 |
| × Aerides × Ascocentrum × Renanthera = Robinara | 1 |
| × Aerides × Ascocentrum × Rhynchostylis = Ronnyara | 6 |
| × Aerides × Luisia = Aeridovanisia* | 0 |
| × Aerides × Neofinetia = Vandofinides | 2 |
| × Aerides × Neofinetia × Rhynchostylis = Sanjumeara | 1 |
| × Aerides × Phalaenopsis = Phalaerianda | 1 |
| × Aerides × Renanthera = Nobleara | 3 |
| × Aerides × Rhynchostylis = Perreiraara | 4 |
| × Aerides × Vandopsis = Maccoyara | 1 |
| × Arachnis = Aranda | 241 |
| × Arachnis × Ascocentrum = Mokara | 95 |
| × Arachnis × Ascocentrum × Phalaenopsis = Bokchoonara | 2 |
| × Arachnis × Ascocentrum × Phalaenopsis × Vandopsis = Sutingara | 1 |

---

*Indicates that the combination has been registered but with a terete form of *"Vanda,"* and none with a strap-leaf *Vanda*. See *Sander's List of Orchid Hybrids*.

219

*Vanda* REGISTRATIONS

| | |
|---|---|
| × Arachnis × Ascocentrum × Renanthera = Yusofara | 4 |
| × Arachnis × Ascocentrum × Rhynchostylis = Bovornara | 1 |
| × Arachnis × Ascocentrum × Vandopsis = Alphonsoara | 1 |
| × Arachnis × Phalaenopsis = Trevorara | 8 |
| × Arachnis × Phalaenopsis × Renanthera × Vandopsis = Macekara | 1 |
| × Arachnis × Renanthera = Holttumara | 28 |
| × Arachnis × Renanthera × Trichoglottis = Andrewara | 1 |
| × Arachnis × Renanthera × Vandopsis = Teohara | 2 |
| × Arachnis × Rhynchostylis = Ramasamyara | 1 |
| × Arachnis × Trichoglottis = Ridleyara | 1 |
| × Arachnis × Vandopsis = Leeara | 2 |
| × Ascocentrum × Ascoglossum × Renanthera = Shigeuraara | 3 |
| × Ascocentrum × Doritis = Ascovandoritis | 4 |
| × Ascocentrum × Doritis × Phalaenopsis = Vandewegheara | 13 |
| × Ascocentrum × Doritis × Phalaenopsis × Renanthera = Paula | 2 |
| × Ascocentrum × Gastrochilus = Eastonara | 1 |
| × Ascocentrum × Luisia = Debruyneara | 4 |
| × Ascocentrum × Luisia × Rhynchostylis = Pageara | 1 |
| × Ascocentrum × Neofinetia = Nakamotoara | 15 |
| × Ascocentrum × Neofinetia × Renanthera × Rhynchostylis = Knudsonara | 1 |
| × Ascocentrum × Neofinetia × Rhynchostylis = Darwinara | 6 |
| × Ascocentrum × Phalaenopsis = Devereuxara | 41 |
| × Ascocentrum × Phalaenopsis × Renanthera = Stamariaara | 6 |
| × Ascocentrum × Phalaenopsis × Rhynchostylis = Himoriara | 3 |
| × Ascocentrum × Renanthera = Kagawara | 64 |
| × Ascocentrum × Renanthera × Rhynchostylis = Okaara | 2 |
| × Ascocentrum × Renanthera × Vandopsis = Onoara | 1 |
| × Ascocentrum × Rhynchostylis = Vascostylis | 112 |
| × Ascocentrum × Rhynchostylis × Vandopsis = Knappara* | 0 |
| × Ascocentrum × Trichoglottis = Fujioara | 1 |
| × Ascocentrum × Vandopsis = Wilkinsara | 15 |
| × Ascoglossum = Vanglossum | 2 |
| × Ascoglossum × Renanthera = Pantapaara | 3 |
| × Doritis = Vandoritis | 4 |
| × Doritis × Phalaenopsis = Hagerara | 4 |
| × Luisia = Luisanda | 11 |
| × Luisia × Neofinetia = Luivanetia | 1 |
| × Luisia × Rhynchostylis = Goffara* | 0 |
| × Neofinetia = Vandofinetia | 14 |

*Indicates that the combination has been registered but with a terete form of "*Vanda,*" and none with a strap-leaf *Vanda.* See *Sander's List of Orchid Hybrids.*

## Vanda

| | REGISTRATIONS |
|---|---|
| × Neofinetia × Renanthera = Renafinanda | 2 |
| × Neofinetia × Rhynchostylis = Yonezawaara | 2 |
| × Phalaenopsis = Vandaenopsis | 104 |
| × Phalaenopsis × Renanthera = Miorara | 7 |
| × Phalaenopsis × Rhynchostylis = Yapara | 3 |
| × Renanthera = Renantanda | 170 |
| × Renanthera × Rhynchostylis = Joannara | 5 |
| × Renanthera × Trichoglottis = Raganara | 1 |
| × Renanthera × Vandopsis = Hawaiiara | 14 |
| × Rhynchostylis = Rhynchovanda | 70 |
| × Rhynchostylis × Vandopsis = Charlieara | 2 |
| × Sarchochilus = Sarcovanda | 1 |
| × Trichoglottis = Trichovanda | 12 |
| × Vandopsis = Opsisanda | 59 |

# Bibliography

American Orchid Society. 1970–present. *Awards Quarterly*. West Palm Beach, FL.

————. 1991. *Handbook on Judging and Exhibition*. 9th ed.

————. 1995. *Orchid Pests and Diseases*.

Arditti, J. 1992. *Fundamentals of Orchid Biology*. New York: Wiley.

Arditti, J., and M. H. Fisch. 1977. Anthocyanins of the *Orchidaceae*: distribution, heredity, functions, syntheses and location. In *Orchid Biology: Reviews and Perspectives*, vol. 1. ed. J. Arditti. Ithaca, NY: Cornell University Press. 117–155.

Bechtel, H., P. Cribb, and E. Launert. 1981. *The Manual of Cultivated Orchid Species*. Cambridge, MA: M.I.T. Press. 3rd ed. 1992.

Brummitt, R. K., and C. E. Powell, ed. 1992. *Authors of Plant Names*, Kew: Royal Botanic Gardens.

Cheng, S., and C. Z. Tang. 1986. A revision of the genus *Vanda (Orchidaceae)* of China. *Acta. Botanica Yunnanica* 8(2):213–221. Translated in 1988 in *Orchid Digest* 52(1):39–44.

Chitanondh, H. 1987. *All-Color Picture Book of Orchids*. Vol. 1. Bangkok: Chitanond.

Christenson, E. A. 1986a. Nomenclatural changes in the *Orchidaceae* subtribe *Sarcanthinae*. *Selbyana* 9:167–70.

————. 1986b. *Ascocentrum* Schltr. *American Orchid Society Bulletin* 55(2):105–111.

————. 1986c. *Dyakia*, a new genus from Borneo (*Orchidaceae: Sarcanthinae*). *Orchid Digest* 50:63–65.

————. 1987. An infrageneric classification of *Holcoglossum* Schltr. *(Orchidaceae: Sarcanthinae)* with a key to the genera of the *Aerides-Vanda* alliance. *Notes from the Royal Botanic Garden Edinburgh* 44(2):249–256.

————. 1992. Notes on Asiatic orchids. *Lindleyana* 7(2):88–94.

————. 1993. Sarcanthine genera, *Aerides*. *American Orchid Society Bulletin* 62(6):594–608.

————. 1994. Nomenclature Notes. *American Orchid Society Bulletin* 63(7):812.

Comber, J. B. 1982a. The genus *Vanda* in Java. *Orchid Digest* 46(4):125–129.

———. 1982b. Sumatran vandas—a new addition. *Orchid Digest* 46(6):204–206.

———. 1990. *The Orchids of Java*. Royal Botanic Gardens, Kew.

Dillon, G. W. 1981. *Beginners Handbook*. American Orchid Society.

Dressler, R. L. 1981. *The Orchids: Natural History and Classification*. Cambridge, MA: Harvard University Press.

———. 1993. *Phylogeny and Classification of the Orchid Family*. Portland, OR: Dioscorides Press.

Engelhard, A. W., J. P. Jones, and S. S. Woltz. 1989. Nutritional factors affecting *Fusarium* wilt incidence and severity. In *Vascular Wilt Diseases of Plants*. Ed. E. Tjamos and C. Beckman. NATO ASI, vol. H28. Berlin: Springer Verlag. Reprinted as Florida Agricultural Experimental Station Journal Series No. 9118.

Engelhard, A. W., and S. S. Woltz. 1976. *Fusarium* wilt of chrysanthemum caused by *Fusarium oxysporum*. AREC Research Report GC 1976-16. Agricultural Research and Education Center. Bradenton, FL: University of Florida.

Firth, K. 1992. Shedding light on your crop. *Greenhouse Grower* July: 58–61.

Fourteenth International Botanical Congress. 1988. *International Code of Botanical Nomenclature*. Königstein, Germany: Koeltz.

Fuchs, R. F. 1987. Trends in *Ascocentrum* breeding. *Orchid Digest* 51(1):165–171.

———. 1990. Vandaceous orchids: their care and culture. *Orchid Digest* 54(1):5–10.

Fukumura, R. 1989. How I started collecting orchids and what I learned. In *Honolulu Orchid Society 50th Anniversary Program*. Honolulu. 6.

Garay, L. 1972. On the systematics of the monopodial orchids I. *Botanical Museum Leaflets* 23(4):149–212.

———. 1974. On the systematics of the monopodial orchids II. *Botanical Museum Leaflets* 23(10):369–375.

———. 1986. Trudelia, a new name for *Vanda alpina*. *Orchid Digest* 50(2):73–77.

Griesbach, R. J. 1983. Orchid flower color—genetic and cultural interaction. *American Orchid Society Bulletin* 52(10):1056–1061.

———. 1985. Polyploidy in orchid improvement. *American Orchid Society Bulletin* 54(12):1445–1451.

———. 1986. That reciprocal cross—is it a mule or hinny? *American Orchid Society Awards Quarterly* 17(3):149–151.

Haager, J. R. 1993. Some new taxa of orchids from southern Vietnam. *Orchid Digest* 57(1):39–44.

Hawkes, R. E. 1965. *Encyclopedia of Cultivated Orchids*. London: Faber and Faber.

Henderson, M. R., and G. H. Addison. 1956. *Malayan Orchid Hybrids*. Singapore: Government Printing Office.

Holttum, R. E. 1972. *Flora of Malaya*. Vol. 1, *Orchids of Malaya*. 3rd ed. (rev). Singapore: Government Printing Office.

International Orchid Commission. 1993. *The Handbook on Orchid Nomenclature and Registration*. 4th ed. London.

Jones, W. 1795. *Asiatic Researches* 4:302.

Kamemoto, H., and R. Sagarik. 1975. *Beautiful Thai Orchids*. Bangkok: Orchid Society of Thailand.

Kamjampai, K. 1990. Diseases and pests of orchids. In 1990 *Catalog of the Kasem Boochoo Nursery*. Bangkok.

Kirch, W. 1967. Commercial aspects of orchid culture. Reproduced in *Do Orchids Grow? And How!* 1990. Ed. D. Woo and W. Nakamoto. Honolulu: Hawaiian Orchid Foundation for the American Orchid Society, Hawaiian Regional Judging Center. 193–197.

Lamb, A. 1983. The scorpion orchids of Sabah. *Orchid Digest* 46(5):175–186.

Mehlquist, G. A. L. 1974. Some aspects of polyploidy in orchids, with particular reference to *Cymbidium, Paphiopedilum,* and the *Cattleya* alliance. In *The Orchids, Scientific Studies*. Ed. C. L. Withner. New York: Wiley. 393–409.

Mirro, M. 1990. For the love of vandas. *American Orchid Society Bulletin* 59(7):690–695.

Northen, R. 1980. *Miniature Orchids*. New York: Van Nostrand Reinhold.

———. 1990. *Home Orchid Growing*. New York: Prentice Hall.

Orchid Society of South East Asia. 1990. New hybrids. *Malayan Orchid Review* 24:13–16. Singapore.

———. 1993. *Orchid Growing in the Tropics*. Singapore: Times Editions.

Postlethwait, J. H., and J. L. Hopson. 1989. *The Nature of Life*. New York: McGraw-Hill.

Pradhan, G. M. 1973. The habitat and growing conditions of *Vanda coerulea,* with cultural hints gleaned therefrom. *Orchid Digest* 37(1):87–91.

———. 1983. *Vanda cristata. American Orchid Society Bulletin* 52(5):464–468.

Pridgeon, A. 1992. *The Illustrated Encyclopedia of Orchids*. Portland, OR: Timber Press.

Quisumbing, E. A. 1981. *The Complete Writings of Dr. Eduardo A. Quisumbing on Philippine Orchids*. 2 vols. Manila: Eugenio Lopez Foundation.

Raven, P. H., R. Evert, and S. Eichhorn. 1992. *Biology of Plants*. New York: Worth.

Sander, F. K. 1906. *Sander's List of Orchid Hybrids*. St. Albans, England. See also *Sander's Complete List of Orchid Hybrids,* 1946, and Addendum: 1946–1960. St. Albans, England. Addenda by the Royal Horticultural Society: 1961–1970; 1971–1975; 1976–1980; 1981–1985; 1986–1990. London.

———. 1927. *Sander's Orchid Guide*. Rev. ed. St. Albans, England.

Schlecter, R. 1913. Die Orchidaceen von Deutsch-Neu-Guinea: *Sarcanthinae. Repertorium Specierum Novarum Regni Vegetabile, Beihefte* 1:953–1039.

Seidenfaden, G. 1988. Orchid genera in Thailand XIV: fifty-nine *Vandoid* genera. *Opera Botanica 95*. Copenhagen.

Seidenfaden, G., and T. Smitinand. 1959–1965. *The Orchids of Thailand: A Preliminary List*. Vols. 1–4. Bangkok: The Siam Society.

Seidenfaden, G., and J. Wood. 1992. *The Orchids of Peninsular Malaysia and Singapore*. Fredensborg: Olsen and Olsen.

Sheehan, T. J. 1990. *American Orchid Society Bulletin* 59(3):277.

Sheehan, T., and M. Sheehan. 1979. *Orchid Genera Illustrated*. New York: Van Nostrand Reinhold.

Smith, J. J. 1905. *Die Orchideen von Java*. Vol. 6, *Flora von Buitenzorg*. Leiden, Netherlands: E. J. Brill.

Sullivan, R. 1993. Obituary of Frederick Campion Steward. *The New York Times*, Sept. 18, 1993. 9.

Suprasert, S. 1975. Notes on *Vanda coerulea* in Thailand. *Orchid Digest* 39(2):51–53.

Sweet, H. R. 1980. *The Genus Phalaenopsis*. Pomona, CA: The Orchid Digest, Inc.

Teo, C. K. H. 1985. *Native Orchids of Peninsula Malaysia*. Singapore: Times Books International.

Teoh, E. S. 1989. *Orchids of Asia*. Singapore: Times Books International.

Valmayor, H. L. 1984. *Orchidiana Philippiniana*. 2 vols. Manila: Eugenio Lopez Foundation.

Veitch, J., and Sons. 1887–1894. *Orchidaceous Plants*. Vol. 2, *Vandeae-Cypripedieae: Sub-tribe Sarcanthinae*. Chelsea, England: H. M. Pollett.

Wallbrunn, H. M. 1984. The art and science of orchid hybridizing. In *Proceedings, Eleventh World Orchid Conference*. Miami, FL. 378–384.

Warne, R. E. 1954. *Vanda sanderiana. Bulletin of the Pacific Orchid Society of Hawaii* 12(3). Reprinted in *Do Orchids Grow? And How!* 1990. Ed. D. Woo and W. Nakamoto. Honolulu: Hawaiian Orchid Foundation for the American Orchid Society, Hawaiian Regional Judging Center. 193–197.

Watson, W. 1890. *Orchids: Their Culture and Management*. London: L. Upcott Gill. Rpt. Princeton, NJ: Darwin Press, 1979.

Williams, B. S. 1894. *The Orchid Growers Manual*. 7th ed. London. Rpt. Codicote, Herts, England: Wheldon and Wesley, 1973.

Withner, C. 1974. *The Orchids: Scientific Studies*. New York: Wiley. 154.

Wright, E., and W. Tippit. 1993. A heavy duty hanger for larger dendrobiums. *American Orchid Society Bulletin* 62 (5):509–510.

Woo, D., and W. Nakamoto, ed. 1990. *Do Orchids Grow? And How!* Honolulu: Hawaiian Orchid Foundation for the American Orchid Society, Hawaiian Regional Judging Center.

# Indices

# Index of Botanical Names

*References to color plates are indicated in italic type, preceding the references to page numbers. The plate number is preceded by the letter P (e.g., P10 indicates Plate 10).*

*Aeranthes,* 100
*Aeridachnis* Bogor, 114
*Aerides,* 34, 93, 95, 99, 109–114, 125, 131
   *Aer. crassifolia,* 109, 111, 112
   *Aer. falcata,* 109
   *Aer. falcata* var. *houlletiana,* 109, 112
   "*Aer.*" *flabellata,* 109, 111, 112
   *Aer. jarckiana. See Aer. leeana*
   *Aer. krabiensis,* 112
   *Aer. lawrenceae,* 109, 110, 112
   *Aer. leeana,* 109, 111
   "*Aer.*" *mitrata,* 120. *See also Seidenfadenia*
   *Aer. multiflora, P70,* 109, 112
   *Aer. multiflora* var. *godefroyae,* 112
   *Aer. odorata,* 109, 111, 112, 113
   *Aer. quinquevulvera,* 109, 111, 112
*Aeridocentrum,* 111
*Aeridovanda,* 110–111
   *Aerdv.* Arnold Sanchez, *P73*
   *Aerdv.* Fuchs Cream Puff, *P71*
   *Aerdv.* Kinnaree, *P72*
   *Aerdv.* Vieng Ping, 110–111, 120
*Angraecum,* 92
*Angraecum eichlerianum,* 92

*Anota violacea. See Rhy. gigantea* ssp. *violacea*
*Arachnis,* 34, 93, 104, 106–107, 113, 121, 125, 132, 145, 160
   *Arach. flos-aeris,* 104–106
     var. *insignis,* 105
   *Arach. hookeriana,* 104–108, 113, 160
     f. *luteola,* 105–106
   *Arach.* Ishbel, 106–107, 108
   *Arach.* Maggie Oei, *P65,* 106
     *Arach.* Maggie Oei 'Red Ribbon', 106–107
   *Arach. maingayi,* 104–106
*Aranda,* 101, 106–108
   *Aranda* Christine, *P66,* 108, 114
   *Aranda* Eric Mekie, *P68*
   *Aranda* Gold Star, 107
   *Aranda* Nancy, *P67*
*Ascocenda,* 87–98, 111, 113, 118–121, 129, 130. *See also Ascocenda* hybrids
   floriferousness of, 81
   flower shape of, 77–78
   origin of, 9, 87, 93
   temperature considerations for, 160

*Ascocenda* hybrids, *P49–P63*
  *Ascda.* Amelita Ramos, 96
  *Ascda.* Baby Blue, 95
  *Ascda.* Bangyikhan Gold
    *Ascda.* Bangyikhan Gold 'Vibroon',
     *P61*
  *Ascda.* Bill Fox, 91, 95
  *Ascda.* Blue Boy, 113
  *Ascda.* Chaisiri, 134
  *Ascda.* Chryse, 94
  *Ascda.* Duang Porn
    *Ascda.* Duang Porn 'Orchidgrove',
     *P62*
  *Ascda.* Elieen Beauty, 119
  *Ascda.* Fuchs Butter Baby, *P63*
  *Ascda.* Fuchs Golden Nugget, 120
  *Ascda.* Fuchs Royal Dragon, 97
    *Ascda.* Fuchs Royal Dragon 'Indigo',
     97
  *Ascda.* John De Biase, 97, 141, 142
    *Ascda.* John De Biase 'Lava Flow',
     *P50*
  *Ascda.* Khun Nok, *P54*
  *Ascda.* Laksi
    *Ascda.* Laksi 'Red Ruby', *P59*
  *Ascda.* Lek, *P55*
  *Ascda.* Meda Arnold, 93, 94, 112, 118,
    119, 124, 142
  *Ascda.* Medasand, 119
  *Ascda.* Memoria Arthur Freed, *P57, P58*
  *Ascda.* Memoria Emily Grove
    *Ascda.* Memoria Emily Grove 'Thai
     Beauty', *P52*
  *Ascda.* Ophelia, 119, 130
  *Ascda.* Peggy Foo, 142
  *Ascda.* Pink Thing, 95
  *Ascda.* Pokai Victory, 96
  *Ascda.* Portia Doolittle, 94
  *Ascda.* Princess Revadee
    *Ascda.* Princess Revadee 'Orchid-
     grove', *P51*
  *Ascda.* Red Gem, 108, 124
  *Ascda.* Rose Seidel, 95
  *Ascda.* Ruth Shave, 130
  *Ascda.* Suk Sumran Beauty, *P56*
  *Ascda.* Tan Chai Beng, 119
  *Ascda.* Tubtim Velvet

  *Ascda.* Tubtim Velvet 'White Angel',
    *P60*
  *Ascda.* Wichot, 95
  *Ascda.* Yip Sum Wah, *P49*, 41, 89, 93,
    94, 96, 97, 112, 119, 130, 134,
    137
    as parent, 141
    remake of, 142
    taxonomy of, 212
*Ascocentrum*, 99, 111, 113, 118, 121,
    129, 130, 131. *See also Ascocen-
    trum* hybrids; *Ascocentrum*
    species
  flower shape of, 93
  genetic influence of, 93, 124, 141
  origin of, 87
*Ascocentrum* hybrids
  *Asctm.* Khem Thai, 92
  *Asctm.* Mona Church, 92
  *Asctm.* Sagarik Gold, 92
  *Asctm.* Sidhi Gold, 92
*Ascocentrum* species
  *Asctm. ampullaceum, P10*, 87, 88,
    89–90, 92, 95
    climatic preferences of, 145
    genetic influence of, 95
    f. *aurantiacum*, 89
    f. *moulmeinense*, 89
  *Asctm. aurantiacum*, 87, 89, 90, 91, 95,
    210
    ssp. *philippinensis*, 88, 210
  *Asctm. aureum*, 87, 91
  *Asctm. christensonianum*, 88
  *Asctm. curvifolium, P11*, 87, 88–89, 90,
    91, 92, 94, 95, 96, 108, 141, 212
    climatic preferences of, 145
    genetic influence of, 141
    temperature considerations for, 160,
    162
  *Asctm. garayi*, 88, 90
  *Asctm. hendersonianum*, 88
  *Asctm. himalaicum*, 87, 91, 92
  *Asctm. insularum*, 88
  *Asctm. miniatum, P12*, 87, 88, 89, 90,
    91, 92, 94, 95, 129
  *Asctm. pumilum*, 87, 89, 91–92
  *Asctm. rubrum*, 87, 88–89

*Asctm. semiteretifolium*, 87–88, 91, 92
*Ascofinetia*, 95, 131
  *Asco.* Furuse, 92
*Asconopsis* Irene Dobkin, 129

*Cattleya*, 72, 93, 137
*Christieara*, 86, 99, 110, 111–113, 160
  *Chtra.* Fuchs Confetti, *P75*, 112
  *Chtra.* Lillian Arnold, 112
  *Chtra.* Manoa 'Robert', 112–113
  *Chtra.* Renee Gerber, *P74*, 112
  *Chtra.* Virginia Braga, 120
*Colchicum autumnale*, 42
*Cymbidium*, 72

*Darwinara*, 131, 211
  *Dar.* Charm, 131
    *Dar.* Charm 'Blue Star', *P97*, 131
*Dendrobium*, 70
  *D.* Pompadour, 103
*Devereuxara*, 129–130
  *Dvra.* Anna Paul, 129
  *Dvra.* Dreamer 'Baby Doll', 130
  *Dvra.* Edith Normoyle 'Norm's Dream', 130
  *Dvra.* Great Expectations, 130
  *Dvra.* Hawaiian Adventure, 129
  *Dvra.* Hawaiian Delight, 129, 130
  *Dvra.* Hawaiian Exotic, *P93*
  *Dvra.* Hawaiian Rainbow, 129–130
*Doritaenopsis* Red Coral, 130
*Doritis*, 130
*Dracula*, 70
*Dyakia*, 88
  *D. hendersoniana*, 88

*Euanthe*, 20, 32, 125

*Holcoglossum*, 20, 103, 104
  *Holcoglossum amesiana*, 215. *See also*
    *V. amesiana*
*Holttumara*, 125
  *Hltmra.* Park Nadesan, *P91*
*Hygrochilus parishii. See Vdps. parishii*

*Kagawara*, 86, 121, 123, 124–125
  temperature considerations for, 160

*Kgw.* Firebird, 124
*Kgw.* William Doi, Jr., 124
*Kgw.* Yoon Weng-Low
  *Kgw.* Yoon Weng-Low 'TOF', *P90*

*Lewisara*, 113–114
  *Lwsra.* Chittivan, 113, 114
  *Lwsra.* Fatima Alsagoff, 114
    *Lwsra.* Fatima Alsagoff 'Zahrah',
      *P76*, 114
  *Lwsra.* Gracia, 113
  *Lwsra.* Max, 113
*Luisa*, 125
*Lutherara*, 124, 126

*Masdevallia*, 72, 92
*Miltonia*, 72
*Miltoniopsis*, 72
*Mokara*, *P69*, 108–109
  *Mkra.* Clark Kuan, 108
  *Mkra.* Khaw Phaik Suan, 108
  *Mkra.* Ooi Leng Sun, 108
  *Mkra.* Redland Sunset
    *Mkra.* Redland Sunset 'Robert's
      Ruby', *P69*

*Nakamotoara*, 131
  *Nkmtra.* Cinderella 'Hawaii', *P96*
*Neofinetia*, 95, 131–132
  *N. falcata*, *P95*, 92, 131
*Neostylis*, 95, 131

*Odontoglossum*, 70, 92
*Opsisanda*, 133
*Opsistylis*, 132

*Paphiopedilum*, 70, 73
*Papilionanthe*, 20, 34, 103, 104, 106,
    107, 125
  climatic preferences of, 145, 160
  *P. hookeriana*, 100, 104
  *P.* Miss Joaquim, 103, 104
  *P. teres*, 100, 103, 104
*Paraphalaenopsis*, 125–128, 129–131,
    172
  *P.* Boediardjo, 127, 129
  *P. denevei*, *P92*, 126, 127

*P. labukensis*, 126
*P. laycockii*, 126, 127, 130
*P. serpentilingua*, 126–127
*P.* Sunny, 127
*Perreiraara*, 120
  *Prra.* Blue Charm, 120
  *Prra.* Luke Thai, 120
    *Prra.* Luke Thai 'Pat Howell', *P86*
  *Prra.* Porchina Blue, 120
*Phalaenopsis*, 19, 70, 82, 93, 103, 107,
    123, 124, 125–131, 133
  *Phal. amabilis*, 129
  *Phal.* Barbara Moler, 129
  *Phal.* Doris, 127, 129
  "*Phal.*" Doris Thornton, 127
  *Phal.* Fairvale, 128
  *Phal. lindenii*, 32
  *Phal.* Norm's Fantasy, 130
  *Phal.* Stewart's Pride, 209
*Pleurothallis*, 70

*Renantanda*, 121, 122–124
  *Ren.* King Crimson, 123
  *Ren.* Memoria Marie Killian, 123
*Renanthera*, 93, 95, 107, 121–125, 126
  *Ren.* Brookie Chandler, 122
  *Ren. coccinea*, 121, 123, 124
  *Ren. imschootiana*, 122, 123
  *Ren.* Kilauea, 124
  *Ren.* King Crimson, *P89*
  *Ren.* Manila, 122
  *Ren.* Memoria Marie Killian
    *Ren.* Memoria Marie Killian 'Eric's
    Red Imp', *P87*
  *Ren. monachica*, 122, 123–124, 125
    *Ren. monachica* 'Suzanne', *P88*
  *Ren. philippinensis*, 122, 123, 124–125
  *Ren. storiei*, 122, 123, 124, 125
*Renanthopsis*, 123–124
  *Rnthps.* Mildred Jameson, 124
*Rhynchostylis*, 93, 95, 114–116, 118, 124,
    126, 131, 132
  *Rhy. coelestis*, 114, 115, 117, 118, 119
    genetic influence of, 120
  *Rhy. coelestis* f. *alba*, *P77*, 115–120
  *Rhy. gigantea*, *P78*, *P79*, 114, 115,
    117, 118, 120

    ssp. *violacea*, 115–116, 117
  *Rhy. retusa*, 114, 116, 118, 120
  *Rhy. violacea. See Rhy. gigantea* ssp.
    *violacea*
*Rhynchovanda*, *P80*, 116–118
  *Rhv.* Blue Angel, 117, 118, 120
  *Rhv.* Galen Kanayama, 117
  *Rhv.* Sagarik Wine, 117
  *Rhv.* Wong Yoke Sim, 117, 118
*Ronnyara*, 120–121
  *Rnya.* Don-Ron Twin, 120
  *Rnya.* Melba Coronado, 120
  *Rnya.* Ronny Low, 120

*Saccolabium*, 88
  *S. micranthum*, 88
*Sarcanthopsis warocqueana. See Vdps.*
  *warocqueana*
*Sarcochilus*, 95
*Seidenfadenia*, 120
  *S. mitrata*, 109
*Smitinandia*, 88
*Spathoglottis*, 107

*Taprobanea*, 21
*Taprobanea spathulata. See Vanda spathu-*
  *lata*
*Trichoglottis*, 132
*Trudelia*, 21
  *T. alpina*, 23

*Vanda. See also Vanda* hybrids; *Vanda*
  species
  floriferousness of, 80
  flower shape of, 74–78
  flower size of, 78–79
  genetic influence of, 124
  origin of, 19
*Vanda* hybrids, *P24*, *P25*, *P29–P47*, *P53*,
  *P64*
  *V.* Antonio Real, *P36*
  *V.* Azur, 58
  *V.* Ben Berliner, 86
  *V.* Boschii, 57, 63
  *V.* Burgeffii, 21, 57, 58, 63
  *V.* Carmen Coll, *P44*
  *V.* Caroline J. Robinson, 57
  *V.* Clara Shipman Fisher, 58

*V.* Deva
   *V.* Deva 'Orchidgrove', *P39*
*V.* Eisensander, 130
*V.* Ellen Noa, 63
*V.* Emily Notley, 58
"*V.*" Emma van Deventer, 104
*V.* Faustii, 58
*V.* Flammerolle, 57, 63
*V.* Frank Scudder, 58
*V.* Fuchs Delight, *P32*, 69–70, 97
*V.* Fuchs Oro, *P29*
*V.* Fuchs Violetta, *P43*
*V.* Gilbert Triboulet, 56, 57, 58, 59, 63
*V.* Golden Doubloon, 55
*V.* Gordon Dillon, *P34, P35*, 44, 69, 70
*V.* Grove's Pleasure
   *V.* Grove's Pleasure 'Orchidgrove',
   *P40*
*V.* Helen Adams, 57
*V.* Herziana, 56, 57
*V.* Hilo Blue, 108, 117
*V.* Jason Robert Fuchs, *P38*
*V.* Jennie Hashimoto, 53
*V.* Jill Walker, 57
*V.* Joan Viggiani, *P25*
   *V.* Joan Viggiani 'Orchidgrove', *P24*
"*V.*" Josephine van Brero, 103, 104
*V.* Kahili Beauty, 57
*V.* Kapoko, 57
*V.* Kasem's Delight, 44, 69, 70, 76, 79,
   96–97, 141
   *V.* Kasem's Delight 'Krachai', *P33, 76*
*V.* Keeree, *P42*
*V.* Kekaseh, 128
*V.* Kultana Gold, 97
*V.* Kupperi, 57
*V.* Lester McCoy, 57
*V.* Loke, 57
*V.* Madame Rattana, 44
*V.* Mandai Amber, 128
*V.* Manila, 57, 63
*V.* Mariannae, 56, 57
*V.* Mary Foster, 57
*V.* Maurine Dalton, 57
*V.* Mem. G. Tanaka, 58
*V.* Mem. T. Iwasaki, 57, 58
*V.* Messneri, 58

*V.* Michael Coronado, 78
   *V.* Michael Coronado 'Bojote', *P41*
*V.* Monacensis, 58
"*V.*" Nellie Morley, *P64*, 104
*V.* Oiseau Bleu, 58
*V.* Onomea, 53
*V.* Orglade's Rosy Dawn, 212
*V.* Paki, 57, 86
*V.* Patricia Lee, *P45*
*V.* Pride of Tjipeganti, 57
*V.* Pukele, 94, 96, 212
*V.* Puna, 58
*V.* Rasri Gold
   *V.* Rasri Gold 'Orchidgrove', *P31*
*V.* Rose Davis, 53
*V.* Rothschildiana, 52, 53, 57, 58, 59,
   62, 63, 74, 91, 95, 117, 118, 129,
   209–210
   *V.* Rothschildiana 'Robert', *P28*
*V.* Sansai Blue, *P19*
*V.* Saphir, 58
*V.* Schoellhornii, 58
*V.* Souvenir de Berthe Jozon, 58
*V.* Tatzeri, 56, 57, 58, 59, 63
*V.* Thong Chai, *P37*
*V.* Wettsteinii, 58
*V.* Wirat, 114
*Vanda* species. *See also Vanda*, synony-
   mized or transferred species;
   *Vanda*, uncertain species
*V. alpina*, 23, 60, 62, 214
*V.* × *amoena*, 35
*V. arbuthnotianum*, 214
*V. arcuata*, 214
*V. amesiana*, 103
*V. bensonii*, 23–24, 60, 62, 68, 96, 209,
   214
*V. bicolor*, 214
*V.* × *boumaniae*, 214
*V. brunnea*, 24, 60, 62, 214
*V. celebica*, 28, 214
*V.* × *charlesworthii*, 24, 58, 212
*V. chlorosantha*, 21, 214
*V. coerulea*, *P1, P13–P19, P48*, 24–25,
   32, 56, 58–63, 65, 67–70, 82, 96,
   113, 128, 129, 133, 214
   climatic preferences of, 25, 145

flower shape of, 76
genetic influence of, 49–53, 77, 82,
    83, 85, 140, 141
latitude of origin, 155
taxonomy of, 209
temperature considerations for, 25,
    160, 162
tetraploid plants of, 43
watering requirements of, 173
*V. coerulea–V. sanderiana* hybrids,
    52–54
*V. coerulea* 'Grove's Delight', *P14*
*V. coerulea* 'Lois Grove', *P13*
*V. coerulea* 'Orchidgrove', *P15*
*V. coerulescens*, 25–26, 60, 62, 95, 96,
    160, 214
*V. concolor*, 26, 60, 62, 64, 214
*V. × confusa*, 214
*V. crassiloba*, 214
*V. cristata*, *P20*, 21, 23, 26, 60, 62, 68,
    86, 96, 128, 129, 131, 214
    temperature considerations for, 160
    *V. cristata* 'Kerry Slora', *P20*
*V. dearei*, *P2*, 26–27, 58, 60, 61, 62,
    64, 68–70, 74, 82, 128, 137, 214
    climatic preferences of, 145
    genetic influence of, 54–55, 85
    temperature considerations for, 160,
        162
*V. denisoniana*, *P3*, 26, 27, 60, 61, 62,
    64, 67–69, 74, 96, 111, 117, 214
    temperature considerations for, 162
    var. *hebraica*, 24, 27
*V. devoogtii*, 214
*V. flavobrunnea*, 214
*V. foetida*, 27, 60, 61, 62, 214
*V. fuscoviridis*, 214
*V. gibbsiae*, 215
*V. griffithii*, 21, 215
*V. hastifera*, 27–28, 215
*V. helvola*, 61, 62, 215
*V. hindsii*, 61, 62, 215
*V. hookeriana*, 100, 103. *See also Papil-
    ionanthe hookeriana*
*V. insignis*, 28, 60, 62, 103, 127, 215
*V. jainii*, 215

*V. javieriae*, 21, 215
*V. kimballiana*, 103
*V. lamellata*, *P4*, *P21*, 28, 60, 61, 62,
    64, 94, 96, 215
    var. *boxallii*, *P21*
    var. *calayana*, 29
    var. *remediosae*, 29
*V. laotica. See V. lilacina*
*V. leucostele*, 215
*V. lilacina*, 29, 60, 62, 215
*V. limbata*, 29, 60, 62, 215
*V. lindeni*, 28, 215
*V. liouvillei*, 30, 60, 61, 62, 128, 215
*V. lombokensis*, 215
*V. luzonica*, *P5*, 30, 53, 58, 59, 60, 61,
    62, 64, 68, 69, 70, 82, 127, 137,
    215
    genetic influence of, 45, 54, 85
    temperature considerations for, 162
*V. merrillii*, *P6*, 31, 53, 60, 61, 62, 63,
    64, 68, 108, 215
    var. *immaculata*, 31
    var. *rotori*, 31
*V. parviflora*, *P22*, 31, 36, 60, 62. *See
    also V. testacea*
    *V. parviflora* 'Luray', *P22*
*V. petersiana*, 215
*V. pumila*, 21, 31, 60, 61, 62, 74, 215
*V. punctata*, 215
*V. roeblingiana*, *P7*, 31–32, 60, 62,
    160, 215
*V. roxburghii*, 19, 35
*V. sanderiana*, 20, 22, 32–34, 56–63,
    65, 67–70, 82, 96, 101, 104, 123,
    127, 128, 129, 215
    climatic preferences of, 32–33, 145
    flower shape of, 74–76
    genetic influence of, 49–53, 83, 84,
        85, 140, 141
    latitude of origin, 155
    taxonomy of, 209
    temperature considerations for, 160,
        162
    watering requirements of, 173
    f. *alba*, 22, 32, 50–51, 140, 210;
        'Eastwind', *P26*

*V. sanderiana* 'Coral Reef', 69

*V. sanderiana* 'Orchidgrove', *frontispiece, P23,* 69

*V. sanderiana* 'Robert', 69

*V. sanderiana–V. coerulea* types, 124

*V. scandens,* 28, 215

*V. spathulata,* 21, 34, 60, 62, 64, 100, 107.

    *V. spathulata* 'Genevieve', *P27*

*V. stangeana,* 34–35, 60, 62, 215

*V. suavis,* 20

*V. subconcolor,* 215

*V. sumatrana,* 35, 60, 62, 215

*V. teres,* 100, 103. *See also Papilionanthe teres*

*V. tessellata, P8,* 19, 35, 58, 60, 62, 64, 68, 96, 215

    genetic influence of, 55

*V. testacea,* 31, 36, 215

*V. thwaitesii,* 215

*V. tricolor, P9,* 20, 30, 36–37, 42, 53, 58, 59, 60, 61, 62, 64, 68, 69, 70, 82, 104, 107, 137, 215

    genetic influence of, 45, 54

    temperature considerations for, 162

    var. *planilabris,* 37

    var. *purpurea,* 37

    var. *suavis,* 36, 129

    *V. tricolor* 'Vieques', 37

*V. vipanii,* 215

*Vanda,* synonymized or transferred species, 215–217

*Vanda,* uncertain species, 217

*Vandaca,* 19

*Vandaenopsis,* 127–129

    *Vdnps.* Jawaii, 127

    *Vdnps.* Laycock Child, 128

    *Vdnps.* Revelation, 128

    *Vdnps.* Revelation 'Jacqueline', 128

    *Vdnps.* Twinkle, 128

*Vandaeranthe* Helmut Paul, 100

*Vandewegheara,* 130–131

    *Vwga.* Hawaiian Flare, *P94*

    *Vwga.* Jerry Vande Weghe, 130–131

    *Vwga.* Tom Raso, 130

*Vandofinetia,* 95, 131

    *Vf.* Virgil, 131

        *Vf.* Virgil 'Botanicals', 131

*Vandofinides,* 131

*Vandopsis,* 93, 121, 132–133

    *Vdps. gigantea, P98,* 132, 133, 134

    *Vdps. lissochiloides,* 132, 133

    *Vdps. parishii,* 132, 133, 134

        var. *marriottiana,* 132–133

    *Vdps.* Sagarik, 134

    *Vdps. warocqueana,* 132, 133

*Vascostylis,* 86, 102, 115, 118–120

    temperature considerations for, 160

    *Vasco.* Blue Fairy, 118

    *Vasco.* Blue Velvet, 119

    *Vasco.* Charles Marden Fitch, *P82*

    *Vasco.* Cynthia Alonso, 119

        *Vasco.* Cynthia Alonso 'Bridget', *P83*

    *Vasco.* Doty, *P84,* 119

    *Vasco.* Five Friendships, *P81*

    *Vasco.* Nong Kham, *P85*

    *Vasco.* Precious, 119

    *Vasco.* Tham Yuen Hae, 131

*Wilkinsara,* 133

    *Wksra.* Golden Delite

        *Wksra.* Golden Delite 'Cynthia', 134

    *Wksra.* Redland Sunrise

        *Wksra.* Redland Sunrise 'Robert', 134

*Yonezawaara,* 131

# Index of Biographical Names

Atherton, Frank C., 57, 127

Banks, Sir Joseph, 35
Berliner, Benjamin C., 98
Boonchoo, Kasem, 13, 44, 69, 79
Brown, Robert, 19

Chaissang, M., 52, 57, 129
Chattalada, Pravit, 13
Chattalada, Revadee, 13
Chitanondh, H., 113
Christenson, Eric A., 14, 21–22, 23, 32,
    34, 60, 87, 88, 89, 90, 91, 92, 100,
    103, 107, 109, 111, 210, 213, 214
Christie, Welda F., 112
Comber, J. B., 30, 31, 35, 37
Coronado, Michael, 120

Dalton, H. K., 57
de Loureiro, Joao, 212
de Saram, Ernest, 57
Devahastin, Charungraks, 13, 66,
    173–174
Dressler, Robert, 22

Fitch, Charles Marden, 14
Fuchs, Robert F., 13–14, 179–180
Fukumara, Roy, 96, 142

Gillmar, S., 58
Gratiot, Jean, 56, 57, 58

Griesbach, Robert J., 14, 42, 59
Grieve, H. G., 106
Grove, Lois, 15

Hawkes, A. D., 125–126
Herz, P., 56, 57
Hesse, Robert H., 14
Holttum, R. E., 37, 88, 107

Jones, Rodney Wilcox, 71–72, 140
Jones, Sir William, 19

Kagawara, H., 124
Kamjaipai, Kulchawee, 200
Kaufman, Richard, 15
Kirsch, Oscar, 63, 117

Lamb, A., 105
Laycock, John, 106
Lewis, Gracia, 113
Lindley, John, 88
Lum, Hon, 120

Mehlquist, Gustav A. L., 38, 42, 138
Mirro, Angela, 14
Mirro, Marilyn, 13–14, 79–80, 180
Miyamoto, M., 117
Mizuta, Richard, 117, 119, 124
Mizuta, Stella, 119
Moir, W. W. G., 63, 117
Mok, C. Y., 108

Moore, Arthur, 157
Morley, Harold, 104

Paul, Helmut, 100, 131
Perreira, Robert, 119, 120
Perreira, Susan, 119

Roxburgh, Dr. William, 35
Rutel, Albert, 14–15

Sagarik, Rapee, 64, 65, 92, 115
Sander, David F., 56
Sander, Fred. K., 55–56, 100
Seidenfaden, G., 36, 88–89, 91
Sheehan, Thomas J., 186
Shimadzu, Prince, 57
Shipman, Herbert, 57, 58, 63
Sideris, C. P., 94
Sidran, Claire, 124
Smiles, William, 14
Smitinand, Tem, 88
Sophonsiri, Treekul, 13

Steward, Frederick Campion, 38–39

Takaki, O. N., 131
Takakura, Francis, 117, 118
Tanaka, B., 57, 58
Tanaka, Richard, 63
Tew, John, 124
Tuvapalangkul, Pricha, 191

Vagner, Richard, 95
Valmayor, H. L., 29, 50, 76
Vande Weghe, Jerry, 130–131

Wallbrunn, Henry M., 54, 124
Warne, Robert, 33, 63
Withner, Carl, 14
Wolcot, John, 47
Wright, E. S., 110

Yamada, M., 113
Yip Sum Wah, 120

# Subject Index

Algae, 140, 198–200

Air circulation, 173–175. *See also* Fans

American Orchid Society (AOS), 21–22, 48–49, 67–74, 91, 93. *See also* Standards of judging
  awarded hybrids
    ascocendas, 95–98
    intergenerics, 113, 119, 128–131, 133–134
    vandas, 67–70
  awards, types of, 72–74
  scoresheet of, 218

Ancestry, assessing, 140–141. *See also* Genetic traits

Bloom time, 23–36, 51–52. *See also* Species, descriptions of

Blooming age, 52, 94, 105, 107, 141

Blooming out, 46

Blooming size, 138

Branching, 85

Buying plants, 137–142

Calcium carbonate equivalent, 187

Cell division, 39–40, 101

Cells, 38–39

Chemicals, precautions with, 177, 197–198

Chiang Mai, 66, 146, 151–155, 160–162

Christenson's list, 21, 56, 103, 214–217

Chromosomes, 38, 39–46, 100, 101–102

Climate, 23, 25, 33, 160

Climatic preferences, 25, 32–34, 145, 160

Colchicine, 42

Cold-sensitivity, 26–27

Color, 22, 50–55, 81–82, 97–98. *See also* Hybrids, descriptions of; Species, descriptions of
  *alba* forms, 45, 50–51, 115
  genetic influence on, 45, 49–55, 82, 93, 96, 98, 105, 140–141
  genetic limitations of, 49–50, 53
  pigment, 22, 45, 81

Community pot, 138, 193–194

Containers, 139, 188–193

Cooling, 165–167. *See also* Temperature

Cultivation, 51, 94, 100–101, 104, 109, 143–175
  assessing health under, 146
  balanced conditions in, 144
  basic requirements, 145–146
  managing stressed plants, 203–204
  outdoor, 167–168
  sanitation, 198
  of seedlings, 193–195

Cut-flower trade, 20, 64, 102–103, 104, 106, 107, 108, 110

Day length, 151–152

Disease, 139, 173, 195–202

DNA, 40

Drenching, 171–172
Drying cycles. *See* Watering

England, role in hybridizing, 57–58
Environmental conditions, 143–175. *See also* Cultivation

Fans, 159, 164, 166, 174–175
    safety considerations of, 167
Fertilizer, 178–184
    nitrogen, 83, 178–179
    phosphorus, 178–179
    potassium, 178–179
    regimens of growers, 178–182
    solubility of, 186–188
    sugar, 183. *See also* Glucose
    trace elements, 178–179
Flask, 107, 193–194
Floral qualities, 9–10, 45, 50–55, 68, 71, 74–85, 93–98. *See also* Color; Flower shape; Hybrids, descriptions of; Species, descriptions of
    floriferousness, 50, 80–81, 93
    pedicel, 50, 83–85, 94
        resupinate, 83–84
    petals
        pinched, 76–77
        rib, thickened, 82
        rolled, 76
        twisted, 42, 53
    size, 78–80
    stalk, 50, 52, 83–85
    texture, sparkling, 81–82, 93, 97–98
Flower shape, 49, 50–51, 74–79, 107. *See also* Hybrids, descriptions of; Species, descriptions of
    cupping, 77
    flatness, 48, 77–78
    genetic influence on, 42, 48–53, 74–79
    spread, 75, 77, 78–79
Form vs. variety, 22, 23, 210
Fragrance, 23, 24, 26, 27, 29, 30, 31, 101, 105, 109, 110
    breeding for, 53, 101
    creosote-scented, 35
    fetid, 27

France, role in hybridizing, 56, 58, 129

Genes, 38–47. *See also* Color; Genetic traits
    color-controlling genes, 44–45, 50–55, 81–82
    importance of superior genes, 47, 70
    regulator genes, 44–45
    structural genes, 44
Genetic traits, 45, 49–55, 77, 82–85, 93, 95, 124, 130, 140–141
    natural selection of, 46–47
    potential, 45–46, 99
    transmission of, 38–47
German Orchid Society, 49, 72
Germany, role in hybridizing, 58–59
Gifts, orchids as, 66
Glucose, 181, 182–183
Gram calorie, 153, 154
Greenhouse culture, 101, 104, 109, 158–159. *See also* Cultivation
    microclimates in, 144–145, 174

Habitat, native. *See* Species, descriptions of
*Handbook on Judging and Exhibition,* 77
Hawaii, role in hybridizing, 48, 58, 63, 67, 71, 86, 96, 100, 102, 106, 110, 117, 119, 142
Heating, 162–165
    insulation over heaters, 165
    placement of equipment, 164–165
    thermostat settings, 162
Horticultural oil, 202
Humidity, 168–171. *See also* Misting; Watering
    measuring of, 170–171
Hybridization, 21–23
    contributions of species, 50–55, 67–70, 140–141
        color, 50–55, 82
        flower shape, 50–53, 74
    current status of, 66–67
    difficulties of, 95, 101–102, 106–108, 125
    first hybrids, 56
    goals of, 10, 48, 53, 64, 102–106, 108–109, 110

history of, 48
hybrids other than primary, 58
intergeneric combinations, 93, 219–221
lack of interest in, 57–58
primary hybrids, 56–65, 92
Hybrids, descriptions of
ascocendas, 93–98
intergeneric hybrids, 103–134
vandas, 56–70

Insects. *See* Pests
Insulation, 163, 165
Interior decoration, 101
International Code of Nomenclature for
Cultivated Plants, 56
International Registration Authority for
Orchid Hybrids, 56, 99

Japan, role in hybridizing, 66, 131
Judging. *See* Judging systems; Standards of
judging
Judging systems, 72–74
generic scales, 72
point scale, 73–74
scoresheets, 72

Langley, 153–154
Leaf tissue analysis, 183–184
Leaves, effect of light on, 159
Light, 147–160. *See also* Shade; Solar
radiation
effect of day length on, 151–153
effect of latitude on, 150–156
excessive, 159
insufficient, 159–160
measuring intensity of, 148–149
optimal conditions of, 147, 149–153
optimizing amount available, 158–159
seasonal patterns of, 147–148
total quantity of, 153–156
Light meter, 148–149
Lip. *See also* Hybrids, descriptions of;
Species, descriptions of
fish-tailed, 30
fringed, 112
hooked, 25

judging criteria of, 78
spur, lack of, 32

Malaysia, role in hybridizing, 63–64, 86,
103, 107–110, 117, 120
Maturity
as a factor in flower shape, 77
as a factor in flower size, 79
Mericlones, 137
Meristems, 39
Misting, 164, 165, 168–170
equipment for, 169–170
Mutations, 45

Natural selection, 46–47
Nomenclature, 207–213. *See also* Taxon-
omy
Novelty hybrids, 55, 86, 101, 140
Nutrition, 176–188. *See also* Fertilizer
flower color, effect on, 181

*Orchid Review,* 52
Outdoor cultivation. *See* Cultivation
Overcrowding, 158
Oxalis, 198

Pesticides, precautions with, 177. *See also*
Pests
Pests, 196–197, 202–203
Ploidy, 22, 40–44, 141–142
diploids, 40
tetraploids, 41–43, 81, 104, 106, 108
use of by Thai breeders, 43–44
triploids, 43, 82
use of colchicine to induce, 42
Pod parent, 46, 59
Pollen parent, 46, 59
Pollination, 39
Polyploids. *See* Ploidy
Popularity, 10, 65, 93, 97, 101
Potting, 139, 188–196
Potting media, 191–192
"Pride of Hawaii," 104

Reduction division, 40
Registration, 56

Remakes, 46, 142
Roots, 139, 168, 172, 188
  protruding, 189–191
Royal Horticultural Society (RHS), 22,
    48–49, 56, 61, 72–74, 90

Safety
  with chemicals, 177, 197–198
  with greenhouse equipment, 167
*Sander's List of Orchid Hybrids*, 29,
    55–57, 61, 99–100, 103, 109–111
Scorpion orchid, 105
Seedling, assessing condition of, 139–140
Shade, 156–158, 165
  outdoors, 167
Sibling crosses, 142
Silicon, 179
Singapore, role in hybridizing, 63–64, 86,
    103, 106, 108, 110, 113–114,
    117, 124, 128
Singapore Botanic Gardens, 103, 106–107
Solar radiation, 153–156. *See also* Light
Species, descriptions of
  *Aerides,* 109
  *Arachnis,* 104–105
  *Ascocentrum,* 89–92
  *Neofinetia,* 131
  *Papilionanthe,* 104
  *Paraphalaenopsis,* 126–127
  *Renanthera,* 121–122
  *Rhynchostylis,* 114–116
  *Vanda,* 22–37
  *Vandopsis,* 132–133
Species ancestry, 49–55, 82, 83. *See also*
    Color; Genetics
  assessing, 140–141
Standards of judging
  ascocendas, 97–98
  intergeneric hybrids, 101
  vandas, 48–49, 68, 71–86
    novelty hybrids, 86
Sterility, 102, 106. *See also* Hybridization,
    difficulties of

Taxonomy, 207–213
  of intergeneric hybrids, 99–100
  synonymous species, 20–21, 215–217
  uncertain species, 217
  of *Vanda,* 19–22, 32, 103
Temperature, 25, 160–168. *See also* Cool-
    ing; Heating
  optimal, 160
  tropical patterns of, 160–162
  wintering outdoors, 163
Terete-leaf, 20, 87, 99–100, 103, 127
Tessellation, 24, 25, 27, 28, 30, 34, 35,
    51, 55. *See also* Species, descrip-
    tions of
Thailand. *See also* Chiang Mai
  climate of, 146
  cultivation of vandaceous orchids in,
    139, 180–181
  role in hybridizing, 13, 52, 64, 65–67,
    76, 79, 86, 97, 101, 103, 110,
    117, 124
  use of tetraploids, 43–44
Tissue culture, 102. *See also* Mericlones
Topping, 195–196
Type-specimen, 22

United States, role in hybridizing, 48, 76,
    79, 86, 124–125. *See also* Ameri-
    can Orchid Society; Hawaii, role
    in hybridizing

Vitamin-hormone additive, 176, 182

Water
  pH of, 185–188
  quality of, 177, 178–179, 184–185
Watering, 168, 171–173. *See also* Humid-
    ity; Misting
  drying cycles during, 172–173
  outdoors, 167
  water temperature for, 172